Michael Hilz

Dissipationless Merging and the Evolution of Early-Type Galaxies

Michael Hilz

Dissipationless Merging and the Evolution of Early-Type Galaxies

Südwestdeutscher Verlag für Hochschulschriften

Impressum / Imprint
Bibliografische Information der Deutschen Nationalbibliothek: Die Deutsche Nationalbibliothek verzeichnet diese Publikation in der Deutschen Nationalbibliografie; detaillierte bibliografische Daten sind im Internet über http://dnb.d-nb.de abrufbar.
Alle in diesem Buch genannten Marken und Produktnamen unterliegen warenzeichen-, marken- oder patentrechtlichem Schutz bzw. sind Warenzeichen oder eingetragene Warenzeichen der jeweiligen Inhaber. Die Wiedergabe von Marken, Produktnamen, Gebrauchsnamen, Handelsnamen, Warenbezeichnungen u.s.w. in diesem Werk berechtigt auch ohne besondere Kennzeichnung nicht zu der Annahme, dass solche Namen im Sinne der Warenzeichen- und Markenschutzgesetzgebung als frei zu betrachten wären und daher von jedermann benutzt werden dürften.

Bibliographic information published by the Deutsche Nationalbibliothek: The Deutsche Nationalbibliothek lists this publication in the Deutsche Nationalbibliografie; detailed bibliographic data are available in the Internet at http://dnb.d-nb.de.
Any brand names and product names mentioned in this book are subject to trademark, brand or patent protection and are trademarks or registered trademarks of their respective holders. The use of brand names, product names, common names, trade names, product descriptions etc. even without a particular marking in this works is in no way to be construed to mean that such names may be regarded as unrestricted in respect of trademark and brand protection legislation and could thus be used by anyone.

Coverbild / Cover image: www.ingimage.com

Verlag / Publisher:
Südwestdeutscher Verlag für Hochschulschriften
ist ein Imprint der / is a trademark of
AV Akademikerverlag GmbH & Co. KG
Heinrich-Böcking-Str. 6-8, 66121 Saarbrücken, Deutschland / Germany
Email: info@svh-verlag.de

Herstellung: siehe letzte Seite /
Printed at: see last page
ISBN: 978-3-8381-3539-7

Zugl. / Approved by: München, LMU, Diss., 2012

Copyright © 2012 AV Akademikerverlag GmbH & Co. KG
Alle Rechte vorbehalten. / All rights reserved. Saarbrücken 2012

"Falls Gott die Welt geschaffen hat, war seine Hauptsorge sicher nicht, sie so zu machen, dass wir sie verstehen können."

Albert Einstein

Zusammenfassung

Elliptische Galaxien sind die größten und schwersten, gravitativ gebundenen Sternensysteme im heutigen Universum und beinhalten ein großen Teil an dunkler Materie innerhalb der sichtbaren, stellaren Komponente. In unserem aktuellen kosmologischen Modell wachsen die Strukturen hierarchisch und elliptische Galaxien bilden sich erst spät. Seit kurzem ist es möglich die Vorgänger heutiger elliptischer Galaxien bei einer Rotverschiebung von $z \sim 2-3$ direkt zu beobachten. Diese waren schon damals sehr schwer, aber sie scheinen um einen Faktor 4-5 kleiner zu sein und ihre projizierte Dichteverteilung ist weniger konzentriert, was man anhand eines sogenannten kleinen 'Sersic index' von $n \sim 2-4$ sehen kann. Die stellaren Populationen ihrer heutigen Ebenbilder deuten darauf hin, daß die Entwicklung der kompakten elliptischen Galaxien nicht auf dissipative Prozesse und die Entstehung neuer Sterne zurückzuführen ist. Das Ziel dieser Arbeit ist es die Entwicklung kompakter elliptischer Galaxien mit der Hilfe von mehr als 80 dissipationslosen (stoßfreien) Verschmelzungssimulationen (Merger) zu erklären. Dafür verwenden wir verschieden anfängliche Masseverhältnisse von 1:1 (Major Merger), 1:5 und 1:10 (Minor Merger). Die Virialgleichungen zeigen, daß Minor Merger zu einer schnelleren Entwicklung führen als Major Merger. Wir erzeugen akkurate Anfangsbedingungen, die die Eigenschaften von elliptischen Galaxien darstellen. Unsere Galaxienmodelle sind sphärisch, isotrop und können verschiedene stellare Dichteverteilungen annehmen. Optional können sich die Galaxien in einem massiven Halo aus dunkler Materie befinden. Es zeigt sich, daß all unsere Modelle im dynamischen Gleichgewicht sind. Betrachtet man die Entwicklung von Major Mergern, sieht man, daß sie proportional mit der Masse wachsen ($r_e \propto M$) und ihre projizierten Dichteverteilungen bei allen Radien zunehmen, weshalb deren Sersic index leicht von 4 auf 6 anwächst. Hier ist der dominante dynamische Prozess die sogenannte 'violent relaxation', die mehr dunkle Materie in das Zentrum mischt und dort das Verhältnis zwischen dunkler und sichtbarer Materie, nach einer Merger Generation, um einen Faktor ~ 1.2 erhöht. Der dynamische Prozess in Minor Mergern wird durch sogenanntes 'stripping' beherrscht. Dabei wachsen die Galaxien stark mit zunehmender Masse an ($r_e \propto M^{\geq 2.1}$) und das Verhältnis von dunkler zu sichtbarer Masse ist für die doppelte stellare Masse um einen Faktor ~ 1.8 höher. Die projizierte Dichte wächst hauptsächlich bei größeren Radien und man erhält Sersic indizes von $n \sim 8-10$. Bemerkenswerter Weise geben nur die Galaxienmodelle mit einem zusätzlichen Halo aus dunkler Materie überzeugende Ergebnisse für alle Minor Merger Szenarien. Das bedeutet, daß dunkle Materie eine sehr wichtige Rolle bei der Entwicklungsgeschichte von kompakten, massive Galaxien spielt. Zusammengefasst zeigen wir, daß dissipationslose Minor Merger in der Lage sind, die Entwicklung von kompakten, elliptischen Galaxien zu erklären, da sie die Größe der Galaxien deutlich erhöhen, mit zusätzlicher Masse höhere Verhältnisse von dunkler zu sichtbarer Materie erzeugen und die Sersic Indizes stark anwachsen lassen.

Summary

Early-type galaxies (ellipticals) are the largest and most massive gravitationally bound stellar systems in the present Universe and contain a significant amount of dark matter within their luminous component. Due to the currently favoured cosmological model, where structures grow hierarchically, these systems assemble late. Recent observations are able to detect directly the progenitors of present day ellipticals at redshifts of $z \sim 2-3$. These are already very massive but they seem to be more compact by a factor 4-5 and have less concentrated surface density profiles, represented by a small Sersic index $n \approx 2-4$. The stellar population of their present day counterparts indicate, that their evolution cannot be driven by dissipation and star formation. The primary goal of this thesis is to investigate a scenario for the evolution of compact, high redshift spheroids using more than 80 dissipationless merger simulations with initial mass ratios of 1:1 (equal-mass), 1:5 and 1:10. Virial expectations have indicated, that minor mergers lead to a more rapid evolution than major mergers. We establish accurate initial conditions, which adequately represent the properties of elliptical galaxies. We setup spheroidal, isotropic galaxies with various density slopes for the stellar bulge, which can optionally be embedded in a dark matter halo. All models are shown to be dynamically stable. Regarding equal-mass mergers, we find that the spheroid's sizes grow proportional to the mass ($r_e \propto M$) and the surface densities grow at all radii, indicated by a weak increase of the Sersic index from 4 to 6. Violent relaxation governs the dynamical merging process and mixes more dark matter particles into the luminous regime. Therefore, the central dark matter fraction increases by a factor of ~ 1.2 after one generation of equal-mass mergers. In minor mergers, stripping of satellites is more important. The size per added mass grows significantly ($r_e \propto M^{\geq 2.1}$) and the final dark matter fractions increase by a factor of ~ 1.8, if the stellar mass is doubled. The surface densities increase predominantly a larger radii, leading to large Sersic indices of $n \sim 8-10$. Remarkably, only the galaxy models including a massive dark matter halo give reasonable results for all minor merger scenarios. This indicates, that dark matter plays a crucial role for the evolution history of compact early-type ellipticals. Altogether we show, that dissipationless minor mergers are able to explain the subsequent evolution of compact early-type galaxies, as they very eciently grow their sizes, yield higher dark matter fractions for more massive galaxies and rapidly increase their Sersic indices.

Meinen Eltern

CONTENTS

1 Motivation — 1

2 Formation and Evolution of Elliptical Galaxies — 3
- 2.1 Elliptical Galaxies — 3
- 2.2 History of merger simulations — 7
  - 2.2.1 First simulations of spherical galaxy mergers — 7
  - 2.2.2 Early high resolution simulations — 7
  - 2.2.3 The first unequal mass mergers — 10
  - 2.2.4 Multiple galaxy mergers — 10
  - 2.2.5 The work of Nipoti et al. — 12
  - 2.2.6 Highly resolved Major Mergers — 13

3 Numerical methods — 15
- 3.1 Numerical N-Body codes — 15
  - 3.1.1 Gravitational Softening — 16
  - 3.1.2 Binary Tree — 18
  - 3.1.3 Oct Tree — 18

4 Creating Initial Galaxy models — 21
- 4.1 One-Component Models — 21
- 4.2 Two-Component Models — 26
- 4.3 Stability Tests — 27
  - 4.3.1 Bulge - Only Models — 27
  - 4.3.2 Bulge + Halo Models with Equal Mass Particles — 29
  - 4.3.3 'Realistic' Bulge + Halo Models — 33

5 Kinematics of Merger simulations — 39
- 5.1 Two-Body relaxation — 39
- 5.2 Dynamical Friction & Tidal Stripping — 42

	5.3 Violent relaxation	44
	5.3.1 Lynden-Bell's approach	45
	5.3.2 Other approaches	46
	5.4 Phase Mixing	47

6 Relaxation and Stripping   49
    6.1 Introduction 51
    6.2 Numerical Methods 53
        6.2.1 Galaxy Models 53
        6.2.2 Model Parameters and Merger Orbits 54
        6.2.3 Simulations and Stability Tests 56
    6.3 Analytic Predictions 59
    6.4 Major Mergers 60
        6.4.1 Violent relaxation 60
        6.4.2 Velocity dispersion 65
        6.4.3 System Evolution 68
    6.5 Minor Mergers 75
        6.5.1 Velocity dispersion 78
        6.5.2 System Evolution 78
    6.6 Summary & Discussion 84

7 Size and Profile Shape Evolution   87
    7.1 Introduction 89
    7.2 Simulations 90
    7.3 Size Evolution 92
    7.4 Evolution of Surface Density 94
    7.5 Profile Shape Evolution 98
    7.6 Dark Matter Fractions 102
    7.7 Discussion and Conclusion 104

8 Conclusion & Outlook   107

List of publications   127

CHAPTER 1

MOTIVATION

In the past decades significant understanding on the early evolution of the Universe has been gained. Shortly after the Big Bang, we can observe primordial density and temperature fluctuations in the cosmic microwave background, which are the starting point of galaxy formation. These small density contrasts are the seeds for the first agglomerations of dark matter, which grow to more massive halos, where the barionic gas can cool and form stars and galaxies (White & Rees, 1978). In the current picture of the CDM model, the further evolution and growth of these first, gas-rich disk galaxies is primarily dominated by merging (Toomre & Toomre, 1972). In their hypothesis, Toomre (1977) coined the idea, that major disk mergers may result in intermediate elliptical galaxies (Barnes, 1992; Naab & Burkert, 2003; Naab & Ostriker, 2009). Recent observations have shown, that some of this early-type ellipticals are massive ($M_* \approx 10^{11} M_\odot$), very compact (eective radii of $R_e \sim$ 1kpc) and quiescent at a redshift of $z \sim 2 - 3$ (Daddi et al., 2005; Trujillo et al., 2006; Longhetti et al., 2007; Toft et al., 2007; Zirm et al., 2007; Trujillo et al., 2007; Zirm et al., 2007; Buitrago et al., 2008; van Dokkum et al., 2008; Cimatti et al., 2008; Franx et al., 2008; Saracco et al., 2009; Damjanov et al., 2009; Bezanson et al., 2009).

One major problem of galaxy evolution stems from the fact, that such a population does not exist in the present universe (Trujillo et al., 2009; Taylor et al., 2010). Instead, present day ellipticals are much more extended and their eective radii are larger by a factor of $\sim 4 - 5$. The most promising scenario to pu up a galaxy's size are dissipationless dry major and minor mergers, which are also expected in a cosmological context (Khochfar & Silk, 2006; De Lucia et al., 2006; Guo & White, 2008; Hopkins et al., 2010). As major mergers add a big amount of mass compared to, e.g. the eective size growth or decrease in velocity dispersion, they cannot be the main evolutionary path (White, 1978; Boylan-Kolchin et al., 2005; Nipoti et al., 2009a). Furthermore, they are highly stochastic and some galaxies should have experienced no major merger at all, and would therefore still be compact today. On the other hand, minor mergers can reduce the eective stellar densities, mildly reduce the velocity dispersions, and rapidly increase the sizes by building up extended stellar envelopes, which grow inside-out (Naab et al., 2009; Bezanson et al., 2009; Hopkins et al., 2010; Oser et al., 2010). However, there are doubts whether this scenario works quantitatively (Nipoti et al., 2003, 2009a) or if other physical process are required.

The best way to investigate the process of dissipationless encounters of two or more galaxies are numerical N-body simulations. In recent years the computational power has evolved and increased very quickly, allowing us to perform very high resolution simulations, which significantly reduce the impact of numerical artefacts. Therefore, they are the best way to explore the di cult natu re of mergers, which are highly non-linear phenomena, implying strong potential fluctuations on very short timescales, which violently change the configurations of galaxies. Equipped with powerful numerical tools, we can ask the interesting question, if the new particle distribution, established by a galaxy encounter always gives some universal profile like an isothermal sphere for the stellar component or an NFW-profile (Navarro et al., 1997) for the dark matter halo, as is typically assumed for massive, present-day ellipticals.

A lot of work has already been done in order to push our knowledge of galaxy formation and evolution, but there are still many interesting, open questions, which we want to address in this thesis:

- What processes influence the dynamics of coalescing galaxies?

- Is dissipationless merging a viable mechanism to increase the sizes of compact early-type ellipticals?

- How does the structure change in either a minor or a major merger?

- What is the main driver for the observed inside-out growth of high redshift elliptical galaxies?

In Chapter 2 we start with a short summary of observations concerning the evolution of elliptical galaxies and the previous numerical work before we give an overview of the used N-body codes in Chapter 3. To investigate all the above questions, we develop a program, which is able to create particle distributions of spherical, isotropic systems and check them for stability in Chapter 4. Further, in Chapter 5, we take a closer look at the dynamics of merging galaxies and the involved processes. Our first paper, which will be submitted soon, mainly addressing the investigation of the dynamics and the galaxy evolution is shown in Chapter 6. The eect on observables like the surface density or surface brightness is summarized in Chapter 7, before we finally draw our conclusions in Chapter 8.

CHAPTER 2

OBSERVATIONS

2.1 Elliptical Galaxies

Elliptical Galaxies are the most massive stellar systems in our universe and thought to be the final stage of galaxy evolution. This results from the common picture of galaxy formation and evolution, where structure in the universe grows hierarchically (White & Rees, 1978; Davis et al., 1985). In the favored CDM model (Komatsu et al., 2011), the most massive early-type galaxiess are supposed to be formed in gas rich major disk mergers at a redshift of $z \sim 2-3$ (Davis et al., 1985; Bournaud et al., 2011). Early collisionless simulations of equal-mass disk mergers already showed, that they nicely reproduce the principal structural properties of bright ellipticals (Toomre, 1977; Negroponte & White, 1983; Barnes, 1992), which are slowly rotating systems with shallow central surface brightness profiles (Bender et al., 1989; Kormendy & Bender, 1996; Kormendy et al., 2009; Lauer et al., 2005). Although the formation and evolution of elliptical galaxies strongly depend on the di erent morphologies of the progenitors and encounter geometries, they show a remarkable regularity in their structural properties. The most famous ramification of this regularity is shown in the fundamental plane of elliptical galaxies, which combines their half-light radii r_e, eective surface brightnesses I_e and velocity dispersions   interior to r_e (Djorgovski & Davis, 1987; Faber, 1987; Dressler et al., 1987; Djorgovski et al., 1988; Bender et al., 1992, 1993). It is often explained as

$$R_e \approx {}^a I^b, \qquad (2.1)$$

where observations yield the exponents $a \sim 1.5$ and $b \sim -0.8$, which diers from simple virial expectations, where $a = 2$ and $b = -1$. The reason for this 'tilt' of the fundamental plane is currently not clear, and might be explained by variations in the mass-to light ratio M_*/L or an increase of the central dark matter fraction (Boylan-Kolchin et al., 2005) combined with structural changes (e.g. Capelato et al. 1995;

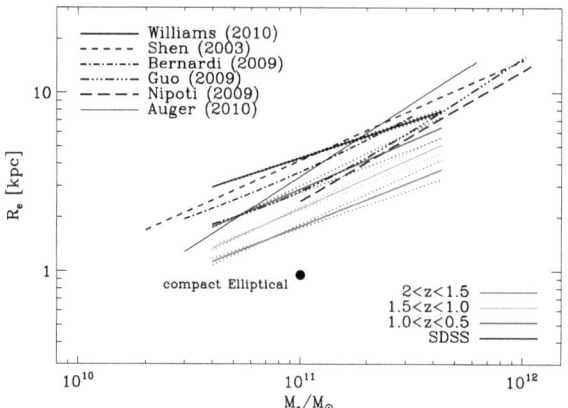

Figure 2.1: This figure shows the position of a compact early-type galaxy (black circle) with respect to the most recent mass-size relations. Due to its extreme compactness, it lies well below the high redshift estimation (red line).

Graham & Colless 1997; Pahre et al. 1998).

Furthermore, all elliptical galaxies are surprisingly well behaved and can all be fitted remarkably well by the Sersic function (Sersic, 1968)

$$I(r) = I_e \cdot 10^{-b_n((r/r_e)^{1/n}-1)}, \qquad (2.2)$$

which is a generalization of the de Vaucouleurs $r^{1/4}$ law. Of course, introducing an additional parameter, the Sersic index n, improves the fit for a big variety of ellipticals, but observational data also supports the idea, that the index n has a physical meaning. For example, it well correlates with the eective radius r_e and the total absolute magnitude of elliptical galaxies (Caon et al., 1993; D'Onofrio et al., 1994; Graham et al., 1996; Graham & Colless, 1997; Graham, 2001; Trujillo et al., 2001, 2002; Ferrarese et al., 2006; Kormendy et al., 2009).

Despite the main body of regular early-type galaxies, recent observations have revealed a population of very compact, massive ($\approx 10^{11} M_\odot$) and quiescent galaxies at z∼2 with sizes of about $R_e \approx 1$kpc (Daddi et al., 2005; Trujillo et al., 2006; Longhetti et al., 2007; Toft et al., 2007; Zirm et al., 2007; Trujillo et al., 2007; Zirm et al., 2007; Buitrago et al., 2008; van Dokkum et al., 2008; Cimatti et al., 2008; Franx et al., 2008; Saracco et al., 2009; Damjanov et al., 2009; Bezanson et al., 2009). Figure 2.1 highlights the position of this population with respect to the most recent mass-size relations (Shen et al., 2003; Bernardi, 2009; Guo & White, 2009; Nipoti et al., 2009a; Auger et al., 2010; Williams et al., 2010). It indicates that present day ellipticals of

2.1 ELLIPTICAL GALAXIES

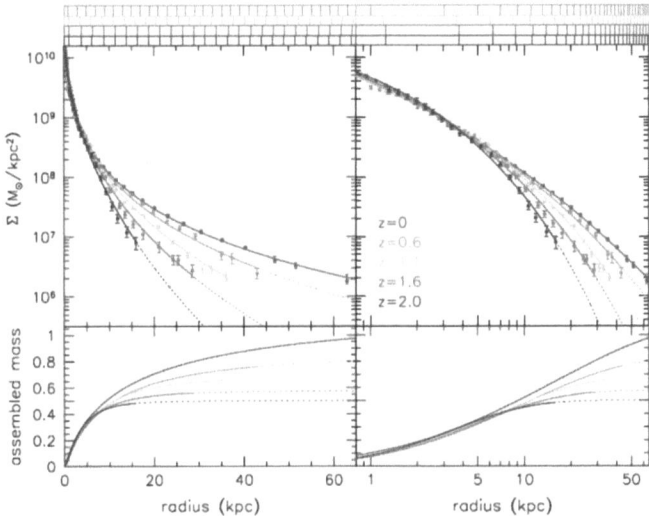

Figure 2.2: The top panels show the observed evolution of the radial surface density of early-type ellipticals from a redshift $z \sim 2$ (blue lines) to the present day (red lines). The bottom panels depict the according mass assembly. Obviously, the central surface densities are not aected and the galaxies grow inside ou t, by developing an outer extended envelope. (Image courtesy of van Dokkum et al. 2010)

similar mass are larger by a factor of 4 - 5 (van der Wel et al., 2008) with at least an order of magnitude lower eective densities and significantly lower velocity dispersions than their high-redshift counterparts (van der Wel et al., 2005, 2008; Cappellari et al., 2009; Cenarro & Trujillo, 2009; van Dokkum et al., 2009; van de Sande et al., 2011). The measured small eective radii are most likely not caused by observational limitations, although the low density material in the outer parts of distant galaxies is di cult to detect (Hopkins et al. 2009a). Their clustering, number densities and core properties indicate that they are probably the progenitors of the most massive ellipticals and Brightest Cluster Galaxies today (Hopkins et al., 2009a; Bezanson et al., 2009).

As this population of early-type galaxies was just found in the last decade, the possible evolution scenarios are under strong debate. However, in a cosmological context, frequent dissipationless galaxy mergers are the most promising scenario to explain the subsequent rapid size growth in the absence of significant additional dissipation and star formation (Cole et al., 2000; Khochfar & Silk, 2006; De Lucia et al., 2006; Guo &

White, 2008; Hopkins et al., 2010). Furthermore, observations and theoretical studies of merger rates support the merger driven evolution, as galaxies undergo, on average, about one major merger since redshift ~ 2 and significantly more minor mergers per unit time (Bell et al., 2006b; Khochfar & Silk, 2006; Bell et al., 2006a; Genel et al., 2008; Lotz et al., 2011). However, using virial estimations (Naab et al., 2009; Bezanson et al., 2009) and the fact that not all galaxies had a major merger since a redshift of $z = 2$, major mergers are not e cient enough to explain such a high size evolution. But they do happen and early theoretical work has shown, that they have a big influence on the structure of spheroidal galaxies (see next section for a summary).

Anyway, recent full cosmological simulations (Khochfar & Silk, 2006; Naab et al., 2009; Oser et al., 2010) and observations (van Dokkum et al., 2010; Williams et al., 2011) pointed out the importance of numerous minor mergers for the assembly of massive galaxies, whose dissipative formation phase is followed by a second phase dominated by stellar accretion (predominantly minor mergers) onto the galaxy. Additionally, minor mergers are particularly e cient in reducing the eective stellar densities, mildly reducing the velocity dispersions, and rapidly increasing the sizes, building up extended stellar envelopes (Naab et al., 2009; Bezanson et al., 2009; Hopkins et al., 2010; Oser et al., 2010, 2011). The latter is also in very good agreement with recent observations of van Dokkum et al. (2010), which indicate, that the central surface densities of early-type galaxies do not change from a redshift of $z \sim 2$, but todays counterparts have assembled a huge amount of mass in the outer parts ($r > 5\mathrm{kpc}$, see also Fig. 2.2).

Although many recent theoretical and observational results indicate, that dissipational minor mergers e ciently boost the size growth of elliptical galaxies, it is yet not clear, if this scenario works quantitatively. Nipoti et al. (2003, 2009a) argue, that dissipationless mergers go in the right direction, but are by far not e cient enough to overcome the big size discrepancy between compact early-types and present day ellipticals. Furthermore, in the first paper (Nipoti et al., 2003) they conclude, that the remnants of multiple mergers neither follow the Faber-Jackson relation (Faber & Jackson, 1976) nor the Kormendy relation (Kormendy, 1977). In the more recent papers (Nipoti et al., 2009b,a) they additionally find that their results introduce a large scatter in the scaling relations of the fundamental plane. The 'tightness' of the fundamental plane sets stringent limitations, so that at maximum 50% of todays ellipticals can have assembled via dry merging (Nipoti et al., 2009a).

Obviously, it is still controversial, if dissipationless mergers are the main evolutionary path for elliptical galaxies. Given the still growing amount of observational data for the high-redshift universe, it is desirable to fill the gap regarding the theoretical background. In this thesis, we want to contribute to the discussion, if dissipational mergers are the driving force, with respect to the evolution of elliptical galaxies or if we need some combinations with other possible scenarios like AGN feedback (Fan et al., 2010).

2.2 History of merger simulations

In this section we give a small overview of the previous work in the field of merger simulations of spheroidal, isotropic galaxy models. As the power of computers increased very fast since the pioneering work in the late 70's, the resolution of the first simulations was really poor, compared to recent ones. Nevertheless, most of the many interesting results are still robust.

2.2.1 First simulations of spherical galaxy mergers

Starting in the late 70's White (1978) made the first N-body simulations of spherical equal-mass mergers, by using only 250 softened particles for each progenitor galaxy (see also Fig. 2.3). One result was, that whenever two galaxies overlap significantly at the pericenter, tidal interactions, mainly dynamical friction, lead to a rapid final coalescence. The final remnants suer from mean field relaxation (violent relaxation), which widens the energy distribution of the binding energies (see Fig. 2.4) and indicates a break in homology. This results in an extended envelope accompanied by a higher central concentration of the final galaxy. Furthermore a strong mixing between 'halo' and central particles occurs during the relaxation process (see also Villumsen 1982), which weakens population gradients during an equal-mass merger (see also White 1980). By a closer investigation of the merger dynamics of radial (head-on) orbits, both progenitor galaxies experience a strong inward impulse during the first overlap, as the mass interior to their position increases immediately. This results in a central contraction relative to the equilibrium configuration, which is followed by a bounce of the particles, when the galaxies separate again and leave the 'deep' potential well. Consequently the outer parts of the galaxies expand and acquire a big amount of the orbital energy (see also van Albada & van Gorkom 1977; Miller & Smith 1980; Villumsen 1982).

In the following work, White (1979) found out, that the density and velocity structure of merger remnants only weakly depend on the initial distribution of the progenitor galaxy and the orbit. The velocity dispersion stays nearly isotropic and the radial density profiles have power-law form $\approx r^{-3}$, which can reasonably well be fitted by a de Vaucoulers surface brightness profile (de Vaucouleurs, 1948).

2.2.2 Early high resolution simulations

Miller & Smith (1980) performed similar simulation, but he was the first, using a very high resolution of nearly 100000 particles. They confirmed the contraction, which occurs just after the closest approach, and find that the initial diameter of the progenitor galaxies decreases by a factor of two, before some particles get lost or build up an extended envelope, in the direction of motion, during the subsequent expansion. Regarding the distribution of binding energies and angular momenta, they also evolve non homologous during the merger event and the escaping particles carry away a large fraction of angular momentum. Furthermore, Miller & Smith (1980) looked at the

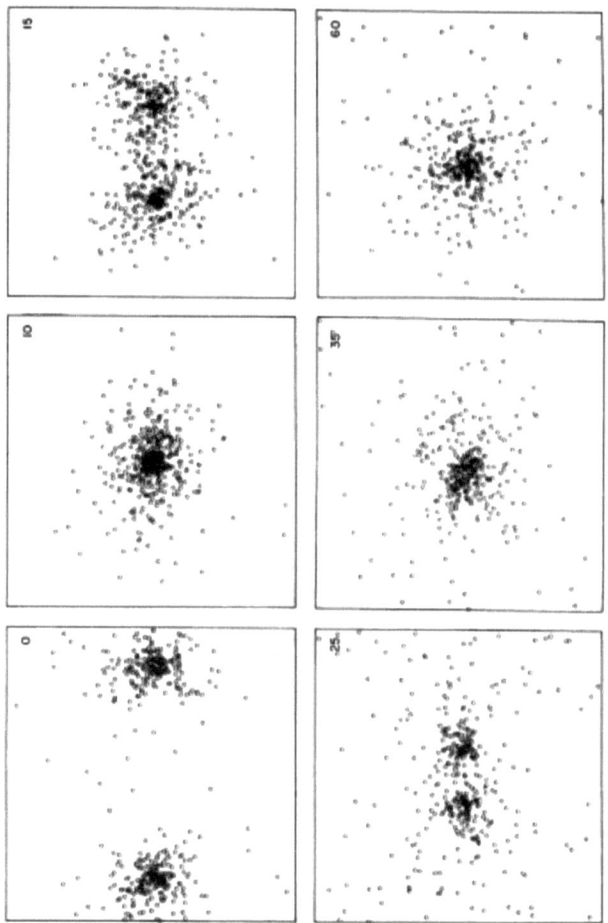

Figure 2.3: This shows one of the first head-on collisions of spherical galaxies from White (1978). Already with this very poor resolution, each galaxies consists of 250 particles, he found very interesting results, regarding the merger dynamics and structural changes of the final remnant.

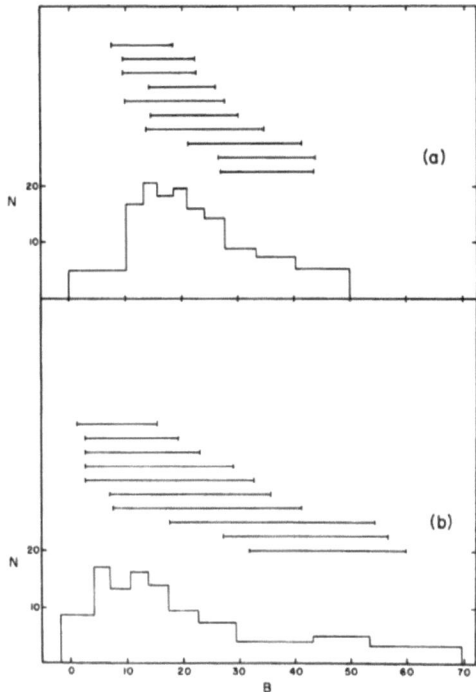

Figure 2.4: In this picture of White (1978) we can already see the eect of violent relaxation, which widens the initial energy distribution (top panel), produces escaping particles (particles with negative energies, bottom panel) and implies a significant amount of mixing, indicated by the width of the bars (see White (1978) for details).

orbits of single particles, during the phase of contraction, and find that all particles are aected as they show a kink in the orbital motion. Finally, due to an energy transfer from the orbit to the galaxies, all remnants pu up and are more loosely bound.

2.2.3 The first unequal mass mergers

Villumsen (1982) was the first who made simulations of both, equal-mass encounters and unequal-mass encounters with mass ratio 1:2. He also claims, that the mixing of the two galaxies is very e cient in the case of equal-mass mergers, which weakens radial metallicity or color gradients, but in the case of unequal-mass mergers this scenario is no longer valid. Because the small in-falling galaxy is less tightly bound it becomes disrupted at an early stage of the merger, and its core would not merge with the one of the host. Especially after the first close encounter, when the particles bounce out of the total combined potential the smaller galaxy explodes and its particles either get lost or assemble in the outer envelope of the bigger host galaxy. Therefore unequal-mass mergers do not weaken the radial gradients, but even might build up a color gradient from the center (older host stars) to the outer parts (blue accreted stars). Furthermore, the remnants of equal-mass mergers either can be prolate, oblate or triaxial, which strongly depends on the orbits angular momentum but all have an anisotropic velocity distribution and their density profiles remains a Hubble profile ($\sim r^{-3}$), which contradicts Lynden-Bell (1967) theory of violent relaxation, which would lead to an isothermal sphere ($\sim r^{-2}$).

2.2.4 Multiple galaxy mergers

Farouki et al. (1983) was the first who simulated higher merger generations with a direct N-body code, starting from a King model. Their particle resolution was lower than some of the previous work, but by a clever sampling for higher generations, the 1000 particles are enough to give interesting results. Assuming energy conservation and homology, they find simple analytic relations for the evolution of equal-mass mergers,

$$= const, \quad R \propto M, \quad _c \propto M^{-2}, \quad (2.3)$$

to which they compared their simulation results. Thereby, they find, that the half-mass radius lies exactly on the relation of Eq. 2.3 but the fraction of the half-mass radius to the radius including 10% of the mass R_h/R_{10} increases with each generation, although it should stay constant, assuming homology arguments (see also Fig. 2.5). Due to the break of homology, they also find a developing low surface brightness envelope in excess of a de Vaucouleurs $r^{1/4}$ law (de Vaucouleurs, 1948). Consequently they find the same core contraction scenario for the remnant as White (1978), which incorporates an increasing central velocity dispersion . As log increases linearly with log M, Farouki et al. (1983) correctly argue, that successive mergers establish a scale-free relation between these properties. By fitting the evolution of the velocity dispersion, they get an exponent $n = 4 - 5$ for $M \propto {}^n$, which nicely agrees with the observed Faber

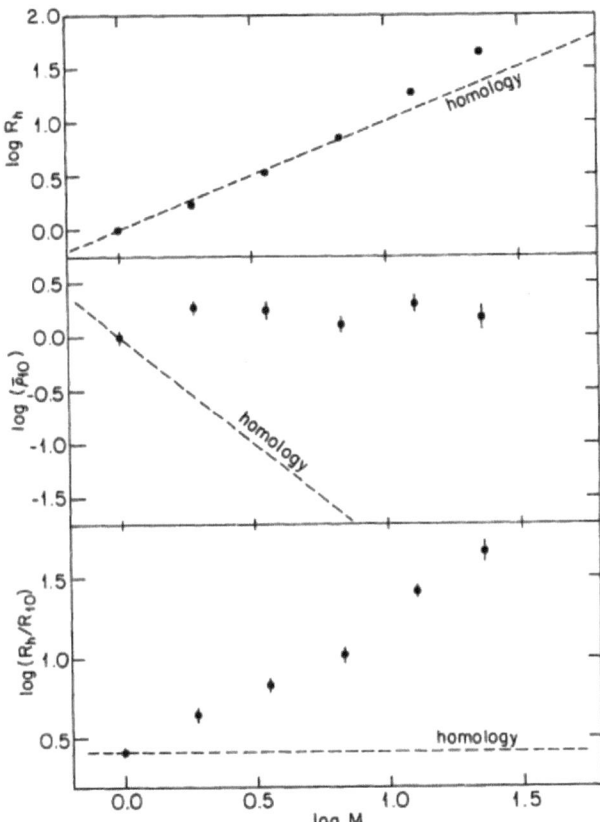

Figure 2.5: This picture of Farouki et al. (1983) indicates nicely the break of homology due to multiple equal-mass mergers. We can see, although the half-mass radius (R_h, top panel) evolves as expected from simple virial expectations, the central densities (middle panel) do not. This is due to a relative contraction of the central regions, as the mass radius including 10% (R_{10}) of the total mass increases much less than the half-mass radius and the ratio R_h/R_{10} grows with each generation.

& Jackson (1976) relation $L \propto$ [4], considering a constant mass-to-light ratio M/L. Furthermore, the velocity dispersion seems to stay isotropic only in the innermost regions, whereas it gets radially biased (to $\approx 50\%$) in the outer parts of the remnant, where a low density envelope has developed.

2.2.5 The work of Nipoti et al.

In Nipoti et al. (2003) they performed hierarchies of equal-mass and unequal-mass mergers. In the end, the final remnants are triaxial systems with axis ratios $0.5 \leq c/a \leq 0.7$ and $0.7 \leq b/a \leq 0.8$, where a,b and c are the major, intermediate and minor axis. By fitting Sersic profiles (Sersic, 1968) to every remnant, they get an increasing Sersic index with increasing mass in accordance with observations, where the more massive ellipticals usually have higher Sersic indices. The velocity dispersion increases with mass and does not stay constant as given by virial expectations for equal-mass mergers. Nipoti et al. (2003) show, that the increase of the velocity can be accounted by the escaping mass, which occurs for each merger generation. However, the half-mass radius evolves like the virial expectations. Traditionally, so far, merger simulations involving a dark matter component have just investigated disk encounters (González-García & van Albada, 2005), thus Nipoti et al. (2003) are among the first who used two-component models for spherical galaxy mergers. Nevertheless, they conclude, that bulges embedded in a dark matter halo, do not give a significant modification in their results. Investigating observable relations, like the fundamental plane and two of its projections (Faber & Jackson 1976- and Kormendy 1977-relation), they find, that although the fundamental plane is well reproduced for their merger hierarchies, the two projections are not.

In a more recent paper Nipoti et al. (2009b) compared a large set of collisionless merger simulations (major and minor) with the fundamental mass plane, which is given by lensing constraints. Thereby, they find that dry merging preserves the nearly isothermal structure of their progenitors and moves galaxies along the mass-plane. But it moves galaxies away from the mass-size and mass-velocity relation, in a way, that the radius increases to rapidly, whereas the velocity dispersion does not. Additionally, dry merging introduces a large amount of scatter in these relations, which sets further constraints on the assembly history and the dark matter fraction within the eective radius increases only because of the rapid size growth and stays constant within a fixed radius. Finally, they conclude that present day early-type galaxies could not have assembled more than 50% of their mass by dry merging.

For the following work, Nipoti et al. (2009a) uses the same simulations and scales his progenitor host to be a compact early-type galaxy with an eective radius of $R_e = 0.9 kpc$, which can be observed at a redshift of $z \sim 2$ (van Dokkum et al., 2009). Considering the dierent major and minor merger hierarchies of the previous paper (Nipoti et al., 2009b), they show, that dry mergers can bring the compact early type galaxies closer to the present scaling relations but quantitatively the process is not e cient enough. Additionally, dry mergers introduce to much scatter to the very tight scaling relations, thus only 45% of the stellar mass of today's early type galaxies can be assembled due to this mechanism.

2.2.6 Highly resolved Major Mergers

Boylan-Kolchin et al. (2005) and Boylan-Kolchin et al. (2006) used highly resolved major merger simulations of two-component models (stellar bulge+dark matter halo) to show, that the fundamental plane is preserved and that the small tilt in the fundamental plane is due to an increasing central dark matter fraction. The latter result is also in good agreement with recent observations which indicate, that stellar mass-to-light ratios are relatively constant with mass and cannot account for the tilt in the fundamental plane. They also pointed out, that the Faber & Jackson (1976) and the mass-size relation strongly depend on the merger orbit, as in-falling galaxies suer much more from dynamical friction for orbits with high angular momentum, which then yields a high energy transfer from the bulge to the halo. The higher the energy transfer, the more compact is the final bulge and the higher becomes the velocity dispersion. On the other hand, by using mainly radial orbits, dissipationless merging is a natural mechanism to change the slopes of the $R - L$ and $L - $ -relation, which can be observed in the brightest cluster galaxies.

CHAPTER 3
NUMERICAL METHODS

3.1 Numerical N-Body codes

Many astronomical objects, such as galaxies, globular and galaxy clusters or especially cosmological cold dark matter systems can be regarded as gravitational N-body systems. In all those systems, the extend of one single body is very small with respect to the spatial distance to other bodies. Then, the interaction of each particle in a gravitating system can simply be described by Newton's law,

$$\mathbf{a}_i = -\sum_{j \neq i} \frac{Gm_j}{r_{ij}^3}(\mathbf{r}_i - \mathbf{r}_j), \tag{3.1}$$

where $\mathbf{a}_i$ is the gravitational acceleration, $\mathbf{r}_i$ and $\mathbf{r}_j$ are the positions of particle i and j, respectively. The particles separation is given by $r_{ij} = |\mathbf{r}_j - \mathbf{r}_i|$, m_j is the mass of particle j and G the gravitational constant.

Although this allows an accurate description of a dynamical system, the computational time for N particles increases proportionally to $\approx N^2$. Therefore, the direct summation or 'Particle-Particle (PP) method' (see also Hockney & Eastwood 1981) is limited to particle numbers of $N \approx 10^5$, which is much too small, compared with recent high-resolution simulations with $\geq 10^{11}$ particles (e.g. the 'Millenium Simulations', Springel et al. 2005; Boylan-Kolchin et al. 2009). These simulations are carried out with a dierent code architecture like a 'hierarchical tree-code', which reduces the computational time to $N \log N$. We use two codes for this thesis, VINE (Wetzstein et al., 2009) and GADGET 3 (which is the updated version of GADGET 2, see Springel 2005), where the first uses a 'binary tree' and the second an 'Oct tree' (Barnes & Hut, 1986). Therefore, we first give a brief summary of the time integration, the force calculation, and the choice of gravitational softening, which is very similar or equal for both codes. Afterwards we show the dierences of the two dierent tree structures.

The equations of motion, according to Newton's law (Eq. 3.1), are ordinary dierential equations,

$$\frac{d\mathbf{r}_i}{dt} = \mathbf{v}_i, \tag{3.2}$$

$$\frac{d\mathbf{v}_i}{dt} = \mathbf{a}_i, \tag{3.3}$$

where $\mathbf{v}_i$ and $\mathbf{r}_i$ are the velocity and the position of particle i, respectively, and the acceleration $\mathbf{a}_i$ is given by Eq. 3.1.

Gravitational forces are long range forces, implying a large dynamical range. Consequently, this aects the equations of motion in a way that they are highly non-linear and cannot be solved analytically if the problem involves more than two bodies. Therefore numerical simulations are the only way to study the formation and evolution of collisionless multi-particle systems. In the numerical approach, the first-order dierential Eqs. 3.2 and 3.3 are replaced by linear dierential equations and the positions $\mathbf{r}_i$ and velocities $\mathbf{v}_i$ are evaluated at discrete time intervals.

Both codes, GADGET and VINE use the common 'leapfrog' integrator to advance the particles in time, but the form is slightly dierent. The explicit leapfrog scheme of VINE is the so-called 'drift-kick-drift' (DKD) method:

$$\mathbf{r}_i^{n+1/2} = \mathbf{r}_i^n + \frac{1}{2}\mathbf{v}_i^n \Delta t_i^n \tag{3.4}$$

$$\mathbf{v}_i^{n+1} = \mathbf{v}_i^n + \mathbf{a}_i^{n+1/2} \Delta t_i^n \tag{3.5}$$

$$\mathbf{r}_i^{n+1} = \mathbf{r}_i^{n+1/2} + \frac{1}{2}\mathbf{v}_i^{n+1} \Delta t_i^n, \tag{3.6}$$

where Δt_i^n is the particle's time step from n to $n+1$. In the 'kick-drift-kick' method used in GADGET, the scheme of the velocities and positions is opposite, in the sense that the positions are updated each integer step and positions each half-integer step. Comparing both schemes, the latter one seems to be slightly more accurate, regarding error properties (Wetzstein et al., 2009).

In order to produce an accurate integration, time steps should be neither too large, nor to small, because too large time steps can destroy the stability of a system and too small time steps waste a huge amount of computational time. Therefore, both codes assign each particle an individual time step, where VINE applies the method of Hernquist & Katz (1989) and the scheme of GADGET is shown in Springel (2005).

3.1.1 Gravitational Softening

One drawback of numerical simulations of astrophysical systems is, that although the underlying physical system like a galaxy with $\sim 10^{11}$ stars, in reality, is collisionless, it is not in numerical simulations. In the latter case, one particle normally represents an

3.1 NUMERICAL N-BODY CODES

aggregate of a large particle number as a simulation is limited to the current hardware (e.g. few times 10^7 particles). Therefore, the evolution time of a numerical system is not smaller than the relaxation time (see also section 4.3) and cannot be treated as a real collisionless system. To overcome this problem, the potential and forces between particles have to be 'softened' in some manner. In practice, the pure Newtonian $1/r$ form of the gravitational potential (Eq. 3.1) and the associated numerical forces at small separations have to be modified by a softening parameter.

There are two common types of gravitational softening in N-body codes, the so-called 'Plummer softening' introduced by Aarseth (1963) and the 'Spline softening'. In the first case, the density function of a single particle is defined as a Plummer sphere, where the force on particle i due to particle j at a distance $r_{ij} = |\mathbf{r}_j - \mathbf{r}_i|$ becomes

$$\mathbf{F}_i = -\frac{Gm_i m_j}{r_{ij}^2 + \epsilon^2} \frac{\mathbf{r}_j - \mathbf{r}_i}{r_{ij}}, \tag{3.7}$$

with the corresponding potential

$$\Phi = -\frac{Gm_j}{(r_{ij}^2 + \epsilon^2)^{1/2}}. \tag{3.8}$$

Here ϵ is the so-called softening length. This implementation is easy and computationally inexpensive, but it never converges completely to the exact Newtonian potential (Eq. 3.1). This choice of softening yields significantly larger force errors compared to the 'Spline softening' (Dehnen, 2001), which we used in both codes.

In this approach, a particle gets smeared out to a finite size and the extended density distribution of the particle is represented by a predefined softening kernel of Monaghan & Lattanzio (1985):

$$W(r_{ij}, h_{ij}) = \frac{\sigma}{h_{ij}^d} \begin{cases} 1 - \frac{3}{2}v^2 + \frac{3}{4}v^3 & \text{if } 0 \leq v < 1 \\ \frac{1}{4}(2-v)^3 & \text{if } 1 \leq v < 2 \\ 0 & \text{otherwise} \end{cases} \tag{3.9}$$

d is the number of dimensions, $v = r_{ij}/h_{ij}$ and σ is the normalization with values of $2/3, 10/(7\pi)$ and $1/\pi$ in one, two and three dimensions, respectively and $h_{ij} = 2.8(\epsilon_i + \epsilon_j)/2$. Then the force is specified as,

$$f_m(r_{ij}) = \frac{4\pi}{m_i} \int_0^{r_{ij}} u^2 \rho(u) du$$
$$= 4\pi \int_0^{r_{ij}} u^2 W(u, h_{ij}) du, \tag{3.10}$$

where the quantity ρ/m_j is replaced by the kernel W. Finally, the force and potential are

$$\mathbf{F}_i = -\frac{G f_m m_i m_j}{r_{ij}^2} \hat{\mathbf{r}}_{ij} \tag{3.11}$$

$$\Phi = -\frac{G f_m m_j}{r_{ij}}. \tag{3.12}$$

Note, that this formulation recovers the exact Newtonian equation for $r_{ij} > 2 \cdot \epsilon_{ij}$ and the force between two particles decreases to zero as $r_{ij} \to 0$.

3.1.2 Binary Tree

The binary tree is constructed bottom-up, where the mutually nearest neighbor particles or particle pairs are replaced by a node. In a first step, imagine that each particle searches for its nearest neighbor, where we require the neighbor to be mutual. Now, consider a system with three particles. If particle B is the nearest neighbor of particle A but the closest neighbor of particle B is C, then B and C are the mutual nearest neighbors and get replaced by a node. The position of the node is its center of mass and its mass is the sum of the particle masses. On the next step, the particles and nodes are again grouped with their nearest neighbor particle or node. Further levels are built accordingly until the last two nodes are combined to the root node and the tree structure is complete. Essential for the construction of such a binary tree is an efficient determination of the nearest neighbors of all particles or nodes for which no nearest neighbor has yet been found. Crucial is also the subsequent combination of these new neighbor pairs into new tree nodes which are then inserted on the next higher level of the tree structure. As one can chose different opening criterions in VINE, we have chosen the same one which is used in GADGET (see next section).

3.1.3 Oct Tree

The oct tree is constructed from top to bottom, as it starts with one initial major cell, which includes all particles. This 'root' cell gets split in 8 cubes of equal size, which are, in the same way, subdivided in smaller subcubes. This process continues until each cube contains only one particle, representing a 'leaf' of the tree, or no particle. A further characteristic of GADGET 3 is, that it only uses monopole terms for the force calculations. Finally, regarding the force calculations on particle i, an acceptance criterion decides whether the force due to a group of other particles at a certain distance is accepted or the cells have to be split up in further cells, ultimately reaching single particles, if appropriate. This criterion controls the introduced errors of the force calculations and the computing time.

The simplest acceptance or so-called cell-opening criterion is usually defined as

$$R_{crit} = \frac{l_j}{\theta} + \epsilon, \qquad (3.13)$$

where ϵ is the particles softening length and l_j the size of the cell. The opening angle θ, ranging from zero to one, defines the minimum distance R_{crit} at which a cell will be accepted for the force calculation or not. GADGET 3 uses a slightly modified criterion

$$\frac{GM_j}{R_{crit}^2}\left(\frac{l_j}{R_{crit}}\right)^2 = \alpha |\mathbf{a}_i^{old}|, \qquad (3.14)$$

where M_j is the mass of cell j and $\mathbf{a}_i^{old}$ is the particles acceleration at the last time step. The advantage of this modification is, that the cell-opening criterion now is adaptive with respect to the system dynamics.

CHAPTER 4

GALAXY MODELS

In this chapter, we describe a way to get stable initial conditions of spherical, isotropic systems, which consist of either a single stellar component or a stellar component embedded in a dark matter halo. One advantage of our program is, that the density slope of the stellar component can be varied and is not fixed for both, a one- and a two-component model. From observations we know, that surface brightness profiles of all kinds of elliptical galaxies are well described by the $R^{1/4}$-law (de Vaucouleurs, 1948) or the more general Sersic $r^{1/n}$ function (Sersic, 1968). Both reproduce global quantities like the eective radius, which is the radius of the isophote enclosing half the total light, and the eective surface brightness. However the derivation of the deprojected three dimensional density distribution and the gravitational potential, which is essential for detailed galaxy modeling is not easily available. One way to overcome this problem is to find analytic density profiles, which resemble in projection the observed surface brightness profiles.

4.1 One-Component Models

The simplest realization of spherical, isotropic galaxies is to create a single sphere of stellar particles. The first two analytic density profiles, resembling the $R^{1/4}$-law, have been proposed by Jae (1983) and Hernquist (1990). They have central stellar densities proportional to r^{-2} and r^{-1}, with central surface densities proportional to R^{-1} and $\ln R^{-1}$, respectively. Dehnen (1993) and Tremaine et al. (1994) independently derived a generalization of these two models,

$$(r) = \frac{(3-\ )M}{4} \frac{a}{r\ (r+a)^{4-}}, \tag{4.1}$$

where a is a scaling radius, M the total mass of the system and defines the slope of the profile. The latter parameter can vary between $0 \leq\ <\ 3$, where $= 1$ and

= 2 represent the Hernquist and Jae model, respectively. The top panel of Fig. 4.1 indicates density distributions of different 's for $M = a = 1$. The central density diverges for all possible slopes except for = 0, where the model resembles a core like structure, i.e. the density becomes constant.

The potential corresponding to Eq. 4.1 is given by Poisson's Equation

$$\Phi(r) = \frac{GM}{a} \times -\frac{1}{2-\gamma}\left[1 - \left(\frac{r}{r+a}\right)^{2-\gamma}\right] \quad \text{for} \quad \gamma \neq 2, \tag{4.2}$$

with the special case of Jae's profile (Jae, 1983),

$$\Phi_2(r) = \frac{GM}{a} \times \ln\frac{r}{r+a} \quad \text{for} \quad \gamma = 2. \tag{4.3}$$

The cumulative mass $M(r)$, half-mass radius $r_{1/2}$ and circular velocity $v_c^2(r)$ are,

$$M(r) = M\left(\frac{r}{r+a}\right)^{3-\gamma}, \tag{4.4}$$

$$r_{1/2,\gamma} = a(2^{\frac{1}{3-\gamma}} - 1)^{-1}, \tag{4.5}$$

$$v_{c,\gamma}^2(r) = \frac{GMr^{2-\gamma}}{(r+a)^{3-\gamma}}. \tag{4.6}$$

Assuming a non-rotating, spherical symmetric system, the radial velocity dispersion is determined by the Jeans equation

$$\frac{1}{\rho}\frac{d}{dr}(\rho \overline{v_r^2}) + 2\beta\frac{\overline{v_r^2}}{r} = -\frac{d\Phi}{dr}, \tag{4.7}$$

where $\beta(r) \equiv 1 - \overline{v_\theta^2}/\overline{v_r^2}$ gives the degree of anisotropy. Later, for simplicity, we only use phase-space distribution functions (DF), which only depend on energy. This implies, that the system has to be isotropic ($\beta(r) = 0$) and as $\overline{v_r^2} = 0$ for $r \to \infty$ we get

$$\overline{v_{r,\gamma}^2}(r) = \frac{1}{\rho(r)}\int_r^\infty \rho\frac{d\Phi}{d}dr, \tag{4.8}$$

which can be solved numerically. In special cases, where 4γ is an integer, Eq. 4.8 has an analytic solution. The radial velocity dispersions show different trends for different density slopes (bottom panel, Fig. 4.1). For $2 < \gamma < 3$ the dispersion diverges towards the center, whereas the models with $0 < \gamma < 2$ converge to zero at the center. In the case of the $\gamma = 0$- and Jae-model ($\gamma = 2$) the central velocity dispersion becomes constant and the latter case resembles a finite isothermal cusp.

4.1 One-Component Models

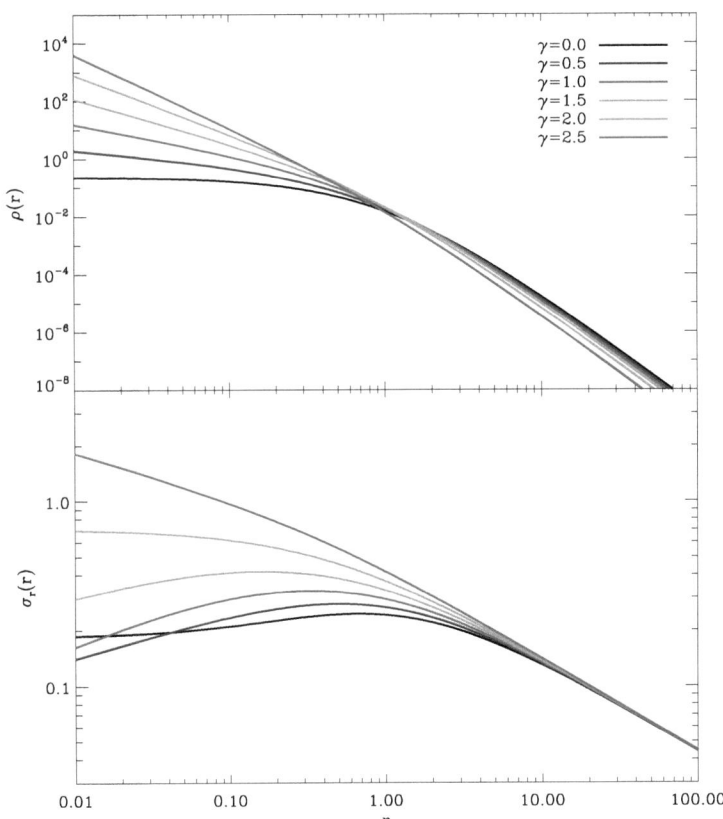

Figure 4.1: Top panel: Density distributions for different Dehnen-Models. For high γ's the profiles are very cuspy and for small ones they become very flat. In the extreme case of $\gamma = 0$ it becomes even constant in the center and resembles a core like structure. Bottom panel: The corresponding radial velocity dispersions show different behavior for different density slopes. Models with $2 < \gamma < 3$ have a diverging central velocity dispersion, whereas those with $0 < \gamma < 2$ converge to zero. There are two special cases, the Jaffe model with $\gamma = 2$, which has a finite isothermal cusp in the center and the $\gamma = 0$ model which becomes constant $\overline{v_r^2} = GM/30a$.

With the density distribution (Eq. 4.1) and the radial velocity dispersion (Eq. 4.8) it is already possible to create a spherical galaxy model, but it is more convenient to use proper distribution functions to get stable initial conditions (Kazantzidis et al., 2004). As we already know the density and the potential, the derivation of the corresponding distribution function $f(\mathbf{r}, \mathbf{v})$ is straightforward. The density of our spherical, isotropic models just depends on the total energy E, thus

$$\rho(r) \equiv \int f(E) d^3\mathbf{v}. \tag{4.9}$$

Inverting this equation with a so called Abel transformation yields the Eddington formula (Eddington, 1916; Binney & Tremaine, 2008), which gives the distribution function for a spherical symmetric density distribution,

$$f(E) = \frac{1}{\sqrt{8}\pi^2}\left[\int_{\Psi=0}^{\Psi=E} \frac{d^2\rho}{d\Psi^2}\frac{d\Psi}{\sqrt{E-\Psi}} + \frac{1}{\sqrt{E}}\left(\frac{d\rho}{d\Psi}\right)_{\Psi=0}\right], \tag{4.10}$$

where the relative potential Ψ and binding energy E are defined, so that $f > 0$ for $E > 0$ and $f = 0$ for $E \leq 0$. The second term on the right hand side of this equation vanishes for any sensible behavior of $\rho(r)$ and $\Psi(r)$ at large radii.

As not all one-component (and no two-component) models, have an analytic expression for $\rho(\Psi)$ we have to transform the integrand of Eq. 4.10 to be a function of radius r,

$$\frac{d^2\rho}{d\Psi^2}d\Psi = \left(\frac{d\Psi}{dr}\right)^{-2}\left[\frac{d^2\rho}{dr^2} - \left(\frac{d\Psi}{dr}\right)\frac{d^2\Psi}{dr^2}\frac{d\rho}{dr}\right]\frac{d\Psi}{dr}dr \tag{4.11}$$

Together with Eqs. 4.1, 4.2 this always results in an analytical expression for the integrand, even for more general γ-profiles (Dehnen, 1993),

$$\left(\frac{d\Psi}{dr}\right)^{-2}\left[\frac{d^2\rho}{dr^2} - \left(\frac{d\Psi}{dr}\right)\frac{d^2\Psi}{dr^2}\frac{d\rho}{dr}\right]\frac{d\Psi}{dr}dr =$$

$$\frac{-2a^3}{(\gamma-2)r(r+a)^3}\left[\gamma\left(\frac{r+a}{r}\right)^2 + 2\gamma\left(\frac{r+a}{r}\right) - \gamma + 4\right]$$

As consequence the integration limits of Eq. 4.10 also have to change, e.g. $\Psi(r) = 0$ corresponds to $r = \infty$ and $\Psi(r) = E$ becomes $r = a/[(1-E)^{\frac{1}{\gamma-2}} - 1]$.

Altogether the DF for the one-component γ-models can be written as,

$$f(E) = \frac{1}{\sqrt{8}\pi^2}\int_{a/[(1-E)^{\frac{1}{\gamma-2}}-1]}^{\infty} \frac{-2a^3}{(\gamma-2)r(r+a)^3} \cdot$$

$$\left[\gamma\left(\frac{r+a}{r}\right)^2 + 2\gamma\left(\frac{r+a}{r}\right) - \gamma + 4\right] \cdot \frac{dr}{\sqrt{(E-\Psi(r))}}, \tag{4.12}$$

4.1 One-Component Models

which can be calculated directly by numerical integration. Alternatively, for all one-component models except $\gamma = 2$, one can use the general solution expressed by Hypergeometric Functions $_2F_1(a, b; c; d)$ (see Abramowitz & Stegun 1970), explicitly given in Baes et al. (2005),

$$f(E, \gamma) = \frac{3-\gamma}{4\pi^3}\sqrt{2E}\left[-(\gamma-4)\,_2F_1\left(1, \frac{\gamma-\gamma}{2-\gamma}; \frac{3}{2}; (2-\gamma)E\right) + \right.$$
$$+ 2(\gamma-3)\,_2F_1\left(1, \frac{1-\gamma}{2-\gamma}; \frac{3}{2}; (2-\gamma)E\right) -$$
$$-2(\gamma-1)\,_2F_1\left(1, \frac{3-\gamma}{2-\gamma}; \frac{3}{2}; (2-\gamma)E\right) +$$
$$\left. +(\gamma)\,_2F_1\left(1, \frac{4-\gamma}{2-\gamma}; \frac{3}{2}; (2-\gamma)E\right)\right]. \qquad (4.13)$$

For all integer or half-integer values of $(2-\gamma)^{-1}$ (e.g. $\gamma = 0, 1, \frac{3}{2}, \frac{7}{4}, \frac{9}{4}, \frac{5}{2}$), all terms of Eq. 4.13 reduce to elementary functions and the distribution function has an analytic solution (Dehnen, 1993). In the particular case of the Jae-model (Jae, 1983) ($\gamma = 2$), the distribution function can best be expressed in terms of the error function and Dawson's integral. For our purpose, we always calculate the DF by numerical integration with high accuracy, thus we get highly stable initial conditions (see section 4.3).

Once the DF has been calculated, we can start to create the particle distributions. First we have to chose the slope of the density profile $\gamma(r)$ and a maximum radius r_{max}, which should be large enough to enclose most of the total system mass. That means, that the cut-off radius should at least be 100 times the scale radius a of the system, which corresponds to the radius enclosing 97, 98 and 99% of the total mass for $\gamma = 0, 1, 2$, respectively (see Eq. 4.4). After specifying the system parameters, we can calculate the gravitational potential $\phi(r)$, before the particles can randomly be sampled from the DF $f(E)$. To establish a particle configuration, we use the acceptance-rejection technique (Kuijken & Dubinski, 1994; Kazantzidis et al., 2004), which works as follows. First we calculate a normalization constant, which is the maximum of the system's phase space

$$const = \left[\left(\frac{r^2}{a^2}\right)\left(\frac{v^2}{v_g^2}\right)f(r, v)\right]_{max}, \qquad (4.14)$$

where a is the system's scale length and v_g the escape velocity at the scale radius. Furthermore, we draw a random number in the interval $[0, 1]$ and if a particle's normalized position in phase space is smaller than this random number, it is accepted, otherwise the particle is rejected and a new particle is sampled.

For simplicity, our initial condition program allows only density slopes $0 \leq \gamma < 2$, but this range already covers most of the observed ranges of stellar density profiles. With $\gamma = 0$, we can create a very flat density distribution with an intrinsic core and for $\gamma \approx 2$, the model has a steep cusp, where the particles are very concentrated in the center. Before we test two one-component models with different density slopes for their stability (Section 4.3.1) we illustrate how to create two-component models, where a stellar bulge is embedded in a dark matter halo.

4.2 Two-Component Models

In Section 2.2, we have seen, that early merger simulations of one-component spheroidal galaxies revealed very interesting results and this models can probably be a good approximation for mergers in centers of clusters, where the dark matter of the approaching satellite galaxy gets stripped very early (González-García & van Albada, 2005). Nevertheless, in the current accepted CDM model, most of a galaxy's mass resides in a dark matter halo, surrounding the stellar bulge. Surprisingly, the dark matter halos seem to have an universal profile, with an inner density slope of r^{-1} and an outer slope of r^{-3}, which is perfectly described by the famous NFW-profile (Navarro et al., 1997)

$$\propto \frac{1}{r(1+r)^2}. \tag{4.15}$$

For simplicity, we chose a Hernquist profile (Hernquist, 1990) for the dark matter distribution, as it is known to resemble the NFW profile in the center and only deviates at larger radii. Then, the density and potential of the halo are

$$\rho_{dm}(r) = \frac{M_{dm}}{2} \frac{a_{dm}}{r(r+a_{dm})^3} \qquad \Phi_{dm}(r) = \frac{GM_{dm}}{r+a_{dm}}, \tag{4.16}$$

where M_{dm} and a_{dm} are the mass and scale radius of the dark matter halo. In the combined system the density distributions of the bulge and the halo are the same, as if you regard the components separately, but the velocities are different. For two-component models, the potential is the sum of the stellar and dark matter potential

$$\begin{aligned}\Phi_T(r) &= \Phi_{dm}(r) + \Phi_*(r) \\ &= -\frac{GM_*}{a_*}\left\{\frac{1}{2-\gamma}\left[1-\left(\frac{r}{r+a_*}\right)^{2-\gamma}\right] - \frac{\mu a_*}{r+a\beta_*}\right\}, \end{aligned} \tag{4.17}$$

where we have introduced two dimensionless parameters $\mu = M_{dm}/M_*$ and $\beta = a_{dm}/a_*$. With the total potential and the density distributions of each component we are able to calculate the distribution functions for the dark matter halo and the stellar bulge. To simplify the calculation of the distribution function, we make Eqs. 4.1, 4.16 and 4.17 dimensionless:

$$\tilde{\rho}_*(r) = \frac{4a_*^3}{M_*} \cdot \rho_*(r) = \frac{(3-\gamma)a_*^4}{r^\gamma(r+a_*)^{4-\gamma}} \tag{4.18}$$

$$\tilde{\rho}_{dm}(r) = \frac{4a_*^3}{M_*} \cdot \rho_{dm}(r) = \frac{2\mu a_*^4 \beta}{r(r+a\beta_*)^3} \tag{4.19}$$

4.3 Stability Tests

$$\tilde{\Phi}_T(r) = -\frac{a_*}{GM_*} \cdot \Phi_T(r) = \frac{1}{2-\gamma}\left[1-\left(\frac{r}{r+a_*}\right)^{2-\gamma}\right] + \frac{\mu a_*}{r+a_*} \quad (4.20)$$

Together with Eq. 4.11 we can calculate the integrands of Eq. 4.10 for both components. Unfortunately, in contrast to the one-component models, the change of the upper integration limit ($\Phi r) = E$ has no analytical solution, so we have to use a numerical minimization routine to solve this equation for r:

$$0 = \tilde{\Phi}(r) - E = \frac{1}{2-\gamma}\left[1-\left(\frac{r}{r+a}\right)^{2-\gamma}\right] + \frac{\mu a}{r+a} - E \quad (4.21)$$

Now the computation of the distribution functions for dierent bulge slopes embedded in a Hernquist dark matter profile is straightforward. First, one has to use Eq. 4.11 to get the derivatives of the densities (Eqs. 4.18, 4.19) and the potetnial (Eq. 4.20), which then get plugged into the Eddington equation (4.10), which gets integrated numerically.

Before sampling the particle distributions of the two components we have to chose a scale length a_* and a mass M_* for the stellar bulge. The scale length and mass of the halo are defined via $\alpha = a_{dm}/a_*$ and $\mu = M_{dm}/M_*$. For the choice of the cuto radii of both components, we have to fulfill the same criteria as before, i.e. they should be large enough to enclose most of the component's mass. After specifying the system properties, the particle distribution is calculated with the acceptance-rejection technique of the previous Section 4.1.

In the next sections, we show some realizations of one- and two-component models and test their stability.

4.3 Stability Tests

Now we test, how the initial conditions of the previous two sections 4.2, 4.1 evolve with time. Using the two N-body codes VINE and GADGET 3, we take dierent galaxy models with varying density distributions for the bulge and dierent particle masses.

4.3.1 Bulge - Only Models

First we look at the one-component models, which represent a stellar bulge without a dark matter component. As we can create dierent density slopes, we take two examples, where one has a shallower core ($\gamma = 0.7$) and the other has a steeper core ($\gamma = 1.4$) than the most popular one of Hernquist (1990). For simplicity, both models have a scale radius of $a_* = 1.0$, a total mass of $M_* = 1.0$ and consist of $N = 5 \cdot 10^5$ particles. The maximum radii of the systems are $r_{sys} = 200$, which are the radii including 98.8 and 99.2% of the total mass for $\gamma = 0.7$ and 1.4, respectively (see also Eq. 4.4). The simulations were performed dimensionless such that the gravitational constant is unity ($G = 1$) and the results can be scaled arbitrary to a preferred unit

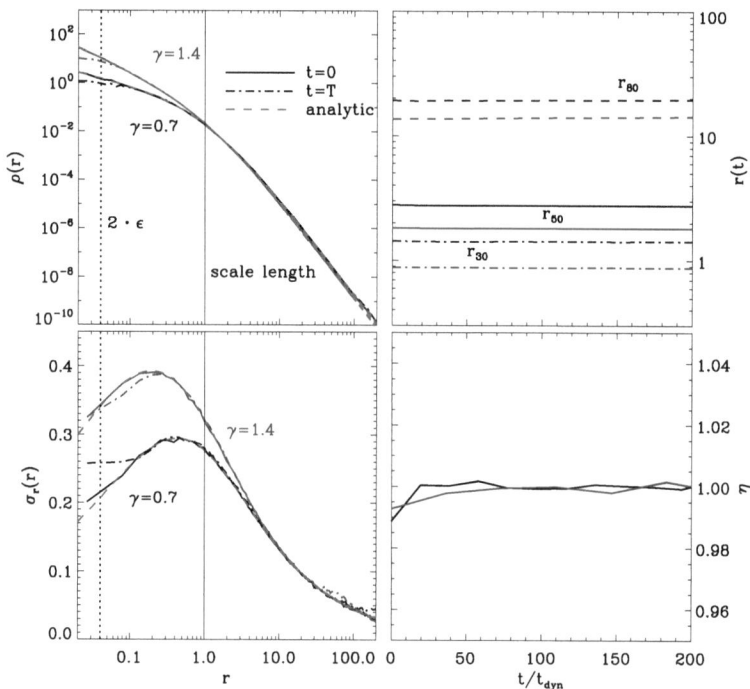

Figure 4.2: Top left panel: The initial (solid lines) radial density profiles stay constant for both, the $= 0.7$ (black) and the $= 1.4$ (blue) model and resemble the analytic profile (red dashed lines) for more than 200 dynamical times. Only inside two softening length $2 \cdot$ (vertical dotted line) the final profiles (dashed-dotted lines) indicate a small decrease, which is due to two-body relaxation. The vertical solid line indicates the scale length of both models. Bottom left panel: Here we illustrate the radial velocity dispersions for both models, which also stay constant over most of the radial range. Only inside 10% of the scale radius, where two-body relaxation becomes prominent, it slightly deviates from the analytical solution. Right panels: The mass radii (top) including 30% (dashed-dotted lines), 50% (solid lines) and 80% (dashed line) of the total mass are perfectly constant for both models and after one or two time steps, the system is in virial equilibrium (see bottom right panel).

4.3 STABILITY TESTS

system. As reference for all stability simulations we use the dynamical time t_{dyn}, which can be regarded as the time a star needs to travel half across a system with a given density. It is defined as

$$t_{dyn} = \sqrt{\frac{3}{16G\bar{\rho}}}, \qquad (4.22)$$

where $\bar{\rho}$ is the mean density within the spherical half-mass radius of the system r_{50} (see also Binney & Tremaine 2008).

In the following we show the stability runs, performed with GADGET 3, but a comparison run with VINE showed the same results. After testing several values, we found the best softening length to be $\epsilon = 0.02$, which gives a good balance between accuracy and computational time.

In the top left panel of Fig. 4.2 we can see, that the density distributions of both, the $\gamma = 0.7$ (black lines) and $\gamma = 1.4$ (blue lines) stay constant for more than 200 dynamical times t_{dyn}. Only within two times the softening length ϵ (vertical dotted line) it slightly decreases, but as the force and potential calculations are not reliable in this regions, we can say that the density distributions perfectly stay constant and agree with the analytic density profiles (red dashed lines). Regarding the radial velocity dispersions of both systems (bottom left panel) we can see that they also show only marginal changes inside 10% of the scale length a_* (vertical solid line). For the flatter $\gamma = 0.7$ density distribution the central deviation is larger, as it contains a factor 5 less particles within $0.1 \cdot a_*$ compared to the more centrally concentrated $\gamma = 1.4$ model. As two-body relaxation strongly depends on the particle numbers (see Section 5.1), and is more efficient for lower particle numbers, shallower density distributions are more affected. For a more detailed description of how two-body relaxation alters our numerical simulations we refer to section 5.1.

The mass radii enclosing 30, 50 and 80% of the total system mass are illustrated in the top right panel of Fig. 4.2. Again they perfectly stay constant over the whole simulation time $t = 200 \cdot t_{dyn}$. In the last panel we can see that the initial galaxy is not perfectly in virial equilibrium as $\eta = 2T/W < 1.0$, but very close. These small deviation is a consequence of the truncation of the system at a radius of $r_{sys} = 200$, which forces the total mass M into a smaller volume as expected. Consequently, the total potential energy W of the system is slightly larger and the kinetic energy T needs little time to adjust. Nevertheless, this effect is negligible, as it has no influence on the densities, the velocity dispersions and the different mass radii and we can conclude, that our scheme to create initial conditions of one-component models with different density slopes yields very good results.

4.3.2 Bulge + Halo Models with Equal Mass Particles

In this section we focus on the stability of two-component models, where a stellar bulge of the previous section is embedded in a more massive dark matter halo. Therefore we first look at three particle configurations where the bulge and dark matter particles all

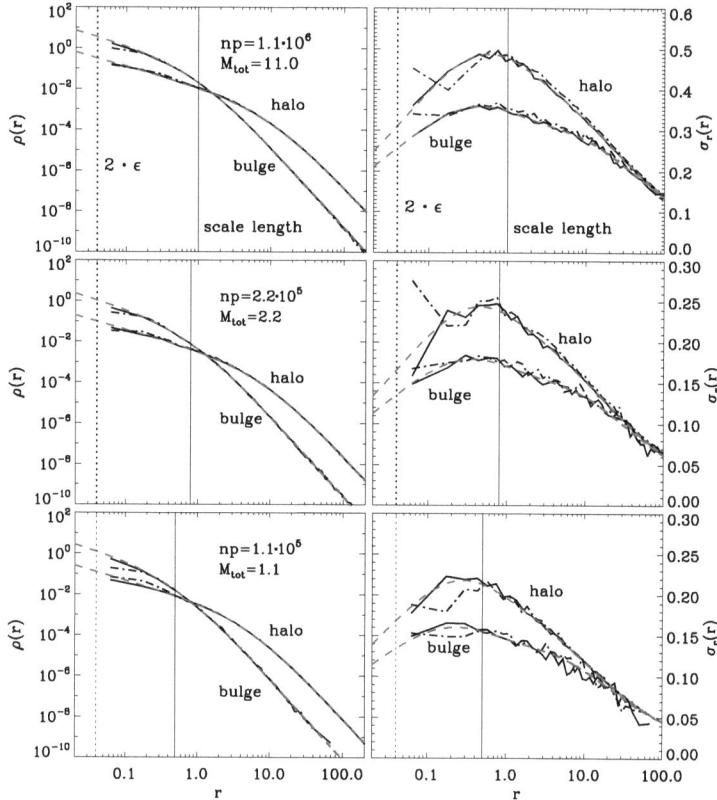

Figure 4.3: The top panels illustrate the densities $\rho(r)$ (left) and radial velocity dispersions $\sigma_r(r)$ (right) for a two-component model of two Hernquist spheres, where a stellar bulge is embedded in a more massive dark matter halo. The total system consists of 10^6 dark matter and 10^5 stellar particles (total particle number $np = 1.1 \cdot 10^6$) and has a bulge mass of $M_{bulge} = 1.0$ and a halo mass of $M_{dm} = 10$. Therefore, all particles have the same mass and we take a force softening length $\epsilon = 0.02$, which gives a good balance, regarding stability and force accuracy. The scale length of the stellar system (vertical solid line) is $a_{bulge} = 1.0$ and the scale radius of the halo is $a_{dm} = 11$. Obviously, the inital (solid lines) and final (dashed dotted lines) density (right panel) and velocity dispersion (left panel) stay constant for 200 dynamical times. The middle and bottom panels show the initial conditions for smaller spheroids with $M_{tot} = 2.2$ and $M_{tot} = 1.1$, scale radii of $a_{bulge} = 0.8$ and $a_{bulge} = 0.5$ and particle numbers of $np = 2.2 \cdot 10^5$ and $np = 1.1 \cdot 10^5$, respectively. The ratios of the masses and scale radii are the same as in the top panel, i.e. $\mu = 10$ and $\eta = 11$. As the particle masses stay the same, we use the same softening which also results in stable initial conditions.

4.3 Stability Tests

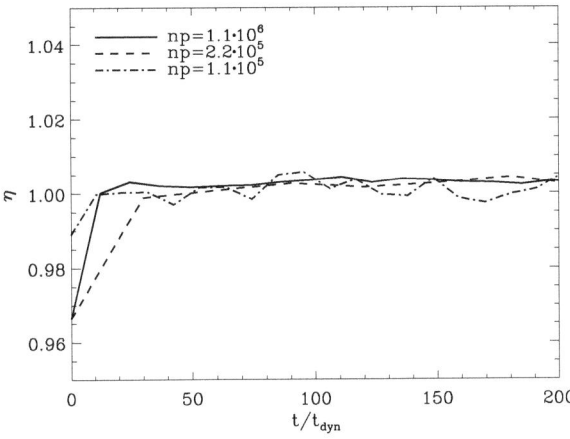

Figure 4.4: This panel illustrates the virial coefficient η for the three models, with $1.1 \cdot 10^6, 2.2 \cdot 10^5$ and $1.1 \cdot 10^5$ particles. Obviously, they are initially not perfectly in virial equilibrium as $\eta < 1$. This results from the truncation of the halo at $r_{sys,dm} \sim 50 \cdot a_{dm}$, which causes a too high initial potential energy W compared to the kinetic energy T.

have the same mass and the stellar bulges represent a Hernquist profile with $\gamma = 1$. As the particle masses do not change, we use a softening length of $\epsilon = 0.02$. Furthermore we keep the ratios for the scale radii $\beta = a_{dm}/a_* = 11$ and masses $\mu = M_{dm}/M_* = 10$ fixed. But all three models have different particle numbers, total masses and scale radii.

The first, most massive, galaxy has a stellar mass of $M_* = 1.0$ and consists of $np = 10^5$ bulge and $np = 10^6$ dark matter particles. We chose a bulge scale length of $a_* = 1.0$ and the cut-off radii are $r_{sys,*} = 200$ and $r_{sys,dm} = 500$ for the bulge and the halo, respectively.

In the top panels of Fig. 4.3, we can see the evolution of the radial density (left) and the radial velocity dispersion (right) of the bulge and the halo. Obviously, the initial conditions (black solid lines), as well as the final profiles after 200 dynamical times (dashed dotted lines) agree perfectly with the analytic profiles (red solid lines) over most of the radial range. Only in the innermost regions the velocity dispersions of the bulge and the halo show some scatter, which again is due to the poor central resolution accompanied by an enhanced two-body relaxation. Furthermore, one can recognize, that the final velocity dispersion profiles are marginally shifted to higher values and the virial coefficient η of the initial model is below unity (see solid line,

Fig. 4.4). This stems from the rather small truncation radius of the halo at $50 \cdot a_{dm}$, which is the radius including 'only' $\sim 96\%$ of the total halo mass. The explanation is the same as for the one-component models, i.e. the total mass is enclosed in a smaller radius, thus the total potential is initially higher and the velocities (or kinetic energy) needs a little time to adjust. However, this eect is negligible, regarding the densities and velocities. Looking at dierent mass radii, we also find that they perfectly stay constant, after a very short phase of slight contraction of less than 3% for the bulge and less than 5% for the bulge's and halo's half-mass radius, respectively. Of course, one can overcome this contraction phase by using much larger cut-o radii for the bulge, but first, the initial variations are very small for our choice and second, to get a perfectly stable two-component model this radius has to be very large, which then increases the computational costs.

In the middle and bottom panels of Fig. 4.3, the galaxies have a stellar mass of $M_* = 0.2$ and $M_* = 0.1$ and consist of $np = 2.2 \cdot 10^5$ and $np = 1.1 \cdot 10^5$ particles, respectively. For both models we chose a scale radius which would lie on the mass-size relation of Fig. 2.1 in chapter 2, i.e. if we scale the previous galaxy to be a compact early-type galaxy (black circle in this figure), than the galaxy with $M_* = 0.2$ would lie on the high redshift relation (red line of Fig. 2.1) for a scale radius of $a_* = 0.8$ and the least massive one for a scale radius of $a_* = 0.5$. We keep the cut-o radii of the massive galaxy, thus $r_{sys,*} = 200$ and $r_{sys,dm} = 500$ for both models and the particle masses also do not change, thus we can take the same softening length of $= 0.02$ for both components.

In Fig. 4.3 the densities and velocity distributions of these models (middle, bottom panels) show very similar results with respect to the more massive, high resolution model (top panels), i.e. the density and velocity profiles are constant over most of the radial range. Especially, the model with $M_* = 0.2$ evolves very close to the $M_* = 1.0$ model, which is not surprising, as these two systems even have nearly the same scale radii. Therefore, their contraction phase is almost identical, which is reflected in the evolution of the virial coecient (Fig. 4.4). In contrast, the model with $M_* = 0.1$ has a much smaller scale radius and the cut-o radius is $r_{sys,dm} \sim 90 \cdot a_{dm}$, thus the truncation radius is the radius containing already $\sim 98\%$ of the total halo mass. Therefore, the initial virial coecient is very close to unity (see also Fig. 4.4). One drawback of the latter model is its comparable small resolution, hence there are only a few particles in the central regions and two-body relaxation is most prominent for this model and causes the relatively high deviations of the final profiles (dashed-dotted lines in the bottom panels of Fig. 4.3) compared to the initial (solid lines) and analytic solutions (red dashed lines).

However, all three models show a high degree of stability, especially in the most relevant regions outside 10% of the bulge's scale radii. If one wants to investigate the very central regions, the resolution has to be very large, which then significantly increases the computation time. Another way to reduce the eect of two-body relaxation in the center would be a slightly larger softening length.

4.3.3 'Realistic' Bulge + Halo Models

Finally we set up initial conditions for early-type galaxies at a redshift of $z \sim 2$, considering observed ratios of the scale radii and masses of the halo and bulge component. To get a proper mass ratio μ we looked at the most recent results of the Halo Occupation Distribution (HOD) models, which determine the link between dark matter halos and the luminous part of galaxies (Moster et al., 2010; Behroozi et al., 2010; Wake et al., 2011). Assuming a luminous mass of $M_* = 10^{11} M_\odot$ the stellar to halo mass ratio of the HOD framework yields values of $M_*/M_{dm} = [0.01, 0.02]$ at redshift $z \sim 2$. Therefore we chose the mean, $M_*/M_{dm} = 0.015$, which then gives $\mu = 66.7$ corresponding to a dark matter halo of $M_{dm} = 6.67 \cdot 10^{12} M_\odot$. Next we have to fix the sizes of both components by chosing proper scale radii.

Applying the mass-size relation of Williams et al. (2010) for the redshift bin $1.5 < z < 2.0$,

$$\log r_e = 0.25 + 0.5(\log(M_*/M_\odot) - 11) \quad [\text{kpc}] \tag{4.23}$$

the eective radius of a $10^{11} M_\odot$ galaxy is $R_e = 1.8$ kpc (see also Fig. 2.1 in chapter 2), which relates to the stellar scale radius a as

$$\frac{R_e}{a} = (2^{\frac{1}{3-}} - 1)^{-1}[0.7549 - 0.00439 + 0.00322^{\ 2} - 0.00182^{\ 3} \pm 0.0007], \tag{4.24}$$

for all $-$models (Dehnen, 1993). For our test models we chose two dierent density slopes, with $= 1.0$ and 1.5. Therefore Eq. 4.24 yields $\frac{R_e}{a} = (1.815, 1.276)$ and the scale radii are $a_* = (1.0, 1.41)$ for $= (1.0, 1.5)$, if we adopt a scaling of $r_{scale} = 1$ kpc. Together with the defined mass scale, where $M_{scale} = 10^{11} M_\odot$ we get the following velocity and time units:

$$v_{scale} = 656 \text{kms}^{-1} \qquad t_{scale} = 1.5 \cdot 10^6 \text{yr} \tag{4.25}$$

This scaling is chosen to describe an early-type galaxy at a redshift of $z \sim 2$, but as the simulations are still dimensionless, one can also use a dierent scaling.

Regarding the size determination of the halo is a little bit more complicated, as we do not use a NFW- but a Hernquist-profile which has a steeper slope at large radii. Therefore we cannot apply the halo concentration c (Bullock et al., 2001; Duy et al. , 2008; Komatsu et al., 2011), which combines the virial radius r_{vir} and the scale radius a_{dm} of the halo profile. Therefore we use a dierent approach, where we set the virial radius of the system equal to the halo mass radius including 80% of the system's total mass M_{80}, thus we can calculate the halo scale radius a_{dm} of the system. To calculate the virial radius r_{vir} we set the virial density (r_{vir}) to 200 times the critical density of the universe $_c$, which yields

$$r_{vir} = \left(\frac{4 M_{vir} G}{225 H(z)}\right)^{1/3}, \tag{4.26}$$

Here, M_{vir} is the virial mass, G is the gravitational constant and $H(z)$ is the time dependent Hubble parameter

$$H(z) = H_0[\Omega_{\Lambda,0} + \Omega_{k,0}(1+z)^2 + \Omega_{m,0}(1+z)^3 + \Omega_{r,0}(1+z)^4]^{1/2} \quad (4.27)$$

with the current matter and radiation density $\Omega_{m,0}/\Omega_{r,0}$, the curvature of the universe $\Omega_{k,0}$ and the cosmological constant $\Omega_{\Lambda,0}$ (see also Mo et al. 2010). In a flat universe $\Omega_{k,0} = 0$ with vanishing radiation density $\Omega_{r,0} \sim 0$ Eq. 4.27 reduces to

$$H(z) = H_0[\Omega_{\Lambda,0} + \Omega_{m,0}(1+z)^3]^{1/2} \quad (4.28)$$

and the Hubble parameter at redshift $z = 2$ is $H(z = 2) = 207$km s^{-1}/Mpc. As the stellar to dark matter mass in the HOD models is defined at the virial radius, we set $M_{vir} = M_* + M_{dm} = 6.76 \cdot 10^{12} M_\odot$ and get a virial radius of $r_{vir} \approx 230$kpc. Next we have to use Eq. 4.4 to calculate the radius including 80% of the total mass of the Hernquist halo $r_{80,h}$. With $\gamma = 1.0$ and $M(r) = 0.8 \cdot M$ we get

$$r_{80,h} = a(4 + \sqrt{20}) \quad (4.29)$$

and for $r_{80,h} = r_{vir}$ the scale radius of the halo is $a_{dm} \approx 27$. Finally, the ratio of the scale radii are $\eta = (27, 19)$ for $\gamma = (1.0, 1.5)$. But as we set $r_{vir} = r_{80,h}$, we have to adopt another mass ratio $\mu = 66.7/0.8 \approx 85$, as μ in the initial condition program gives the stellar to halo mass ratio of the total system. Finally we chose very large cut-off radii of $r_{sys,*} = 200 \cdot a_*$ and $r_{sys,dm} = 100 \cdot a_{dm}$, to limit the contraction effect.

For the stability simulations we use $1.1 \cdot 10^6$ particles for both realizations, which results in more massive dark matter particles ($m_{dm} = 8.5 \cdot m_*$). Therefore the force softening has to be different, i.e. $\epsilon_{dm} = \sqrt{8.5}\epsilon_*$. Furthermore, using different particle masses, two-body relaxation causes mass segregation (see Section 5.1), thus we additionally have to increase both softenings. Finally, we find $\epsilon_* = 0.1$ yields very good results and is still small compared to the effective radius ($\epsilon_* = 0.055 \cdot R_e$).

In the top panels of Fig. 4.5 we illustrate the density profiles of the initial conditions (black solid lines) for $\gamma = 1.0$ (left) and $\gamma = 1.5$ (right), which both stay constant. After 120 dynamical times (dashed dotted lines), the bulge and halo profiles are still in very good agreement with the analytic Hernquist profile (red dashed line). Only inside $2 \cdot \epsilon$ the shallower $\gamma = 1.0$ profile shows minor deviations due to two-body relaxation, which causes the bulge and halo profile to get closer. In the bottom two panels we depict the according evolution of the velocity dispersion profiles, which again show very promising results, as the initial profiles nicely resemble the ones from the Jeans equations (Eq. 4.8) for each component. In the end, they again show only small deviations in the inner parts, where the resolution is lowest and mass segregation, induced by two-body relaxation, is most prominent. But especially the bulge profile is perfectly stable outside 40% of the scale radius.

In the next Fig. 4.6 we can see the evolution of the mass radii including 30, 50 and 80% of the total bulge (top) and halo (bottom) masses. Together with the effective radius R_e (top) and the gravitational radius r_g (bottom), all radii are constant over the

4.3 Stability Tests

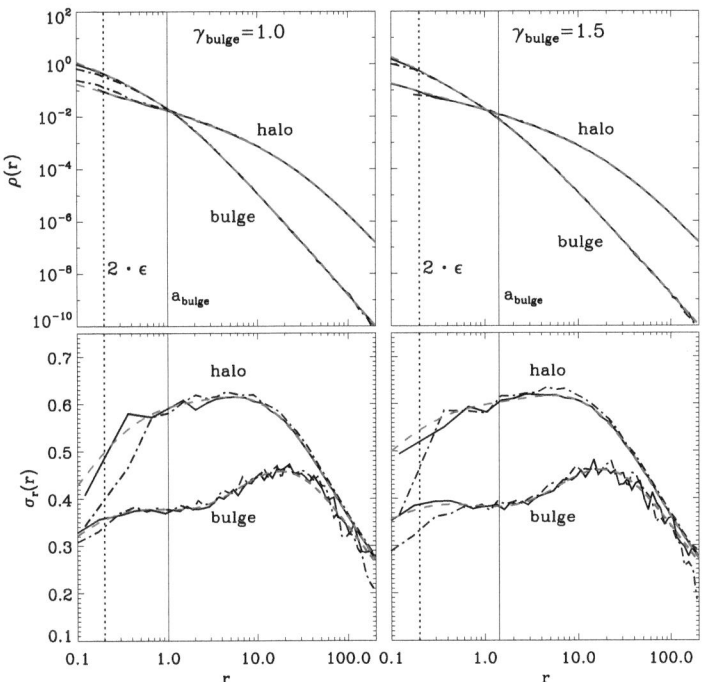

Figure 4.5: The top panels illustrate the densities $\rho(r)$ for the $\gamma = 1.0$ (left) and $\gamma = 1.5$ (right) model with realistic dark to stellar mass ratios. Outside two times the force softening ($\epsilon = 0.055 \cdot R_e$) the densities of both the halo and the bulge resemble perfectly the analytic profile after more than 120 dynamical times. The very small deviations in the central regions are caused by two-body relaxation. The bottom panels illustrate the corresponding radial velocity dispersion profiles, which stay constant over most of the radial range. Only inside 40% the bulge scale radius a they are strongly affected by two-body relaxation.

whole simulation time. Only the innermost bulge radii (r_{30}, top panel) are aected and as the $= 1.0$ profile is shallower than the $= 1.5$ profile, it has less particles in its center and consequently gets slightly more influenced by numerical eects. Therefore its r_{30} finally increases by 7%. However, we conclude that even the 'real' galaxy models are by far stable enough to yield reasonable galaxy models to be used for further applications.

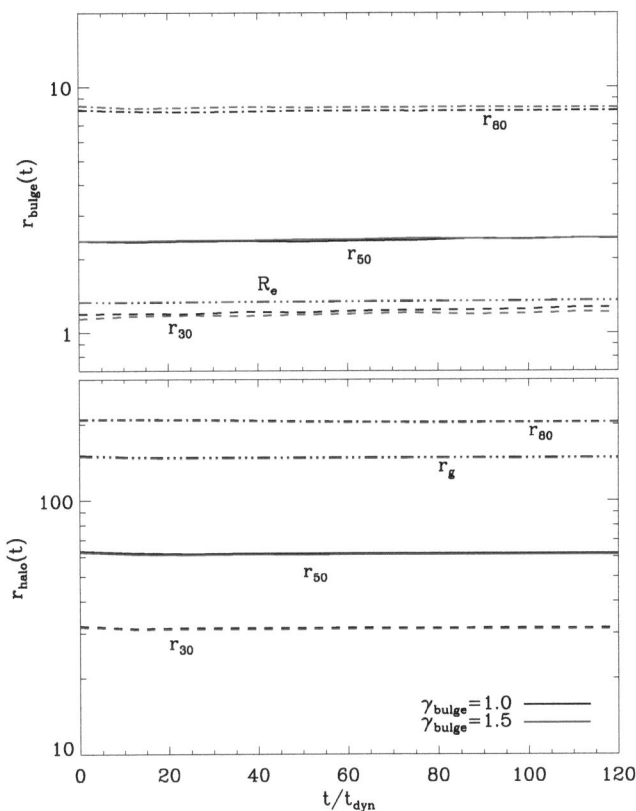

Figure 4.6: The top panel depicts the evolution of the radii enclosing 30,50 and 80% of the total bulge mass for the models with $= 1.0$ (black) and $= 1.5$ (blue). Together with the mean eective radius R_e, they all stay constant for more than 120 dynamical times. Only the innermost radii r_{30} indicate a small increase, which is a little bit larger for $= 1.0$. In the latter shallower model, less particles are in the center and therefore it suers slighlty more from two-body relaxation within the same simulation time (see also Chapter 5). The bottom panel indicates, that all mass radii and the gravitational radius of the corresponding halos stay the same.

CHAPTER 5

KINEMATICS

In the following we want to give a little overview of the dynamical processes, one has to deal with in numerical N-body simulations. All of them have dierent impact for dierent merger scenarios and two-body re laxation strongly depends on the numerical setup of the initial conditions and can be reduced by a clever choice of the gravitational softening length. Dynamical friction and tidal stripping are the dominant processes in minor mergers, whereas violent relaxation is very e cient for major mergers. Although violent relaxation has a strong impact during the final merging process, it rapidly gets dissolved by phase mixing. In contrast to two-body relaxation, all these mechansims are physical and not artificial, thus we first illustrate the influence of two-body relaxation with the help of an test simulation.

5.1 Two-Body relaxation

In the real universe, star and dark matter particles in galaxies, which consist of $N \approx 10^{11}$ stars, are essentially collisionless and feel no perturbation due to a close encounter. However, simulations of isolated galaxies or galaxy mergers, the number of particles is limited by the computational power. Therefore each simulation particle corresponds to a conglomeration of real stars.

The relaxation time t_{relax} denotes the time, when the velocity of one star has changed of the same order as its initial velocity due to two-body encounters (see Binney & Tremaine (2008)). Another useful definition of relaxation time uses the change of the mean square energy compared to the initial mean kinetic energy of a group of particles (Chandrasekhar, 1942). The latter definition yields a relaxation time which is half compared to the first definition.

To quantify the eect of two-body relaxation we chose the first approximation which

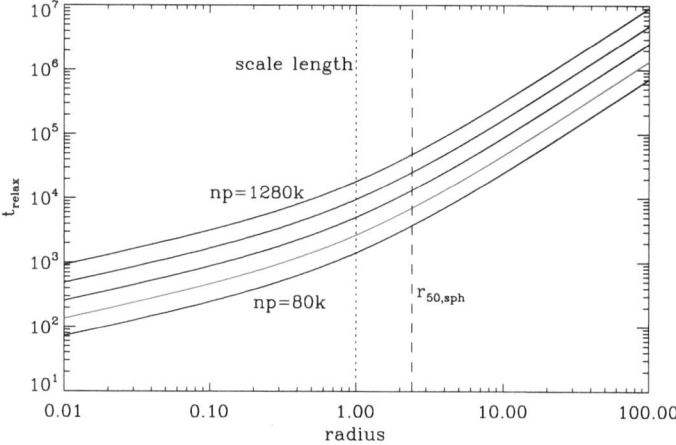

Figure 5.1: The solid lines show the two-body relaxation times of a Hernquist sphere with scale radius $a = 1$ (dotted line) for an increasing number of particles (from bottom to top the particle number increases by a factor of 2). From the lowest particle number ($N = 80000$) to the highest one ($N = 1.28 \cdot 10^6$) the relaxation time at the spherical half-mass radius (dashed line) increases by a factor of ~ 13. The red solid line indicates the particle distribution ($N = 160000$) which we chose for further investigations and for some of our merger simulations in the next sections. For this Hernquist sphere $t_{relax} \sim 7100$ at the spherical half-mass radius.

yields,

$$t_{relax} = \frac{0.1 N}{\ln N} \cdot t_{cross}, \quad (5.1)$$

where N is the particle number. The crossing time $t_{cross} = R/v$ strongly depends on the particles distance R from the center. The velocity v corresponds to the typical velocity of a particle at this radius, which can be approximated by the circular velocity. For a better illustration we chose a one-component Hernquist model, which has a circular velocity $v_c = \sqrt{GMr/(r+a)}$ (see Eq. 4.6). Assuming $G = M = 1$ the crossing time becomes,

$$t_{cross} = \sqrt{r}(r+a), \quad (5.2)$$

where a is the scale radius of the Hernquist sphere. Figure 5.1 shows the radial dependence of the relaxation time for an increasing number of particles. From bottom

5.1 Two-Body relaxation

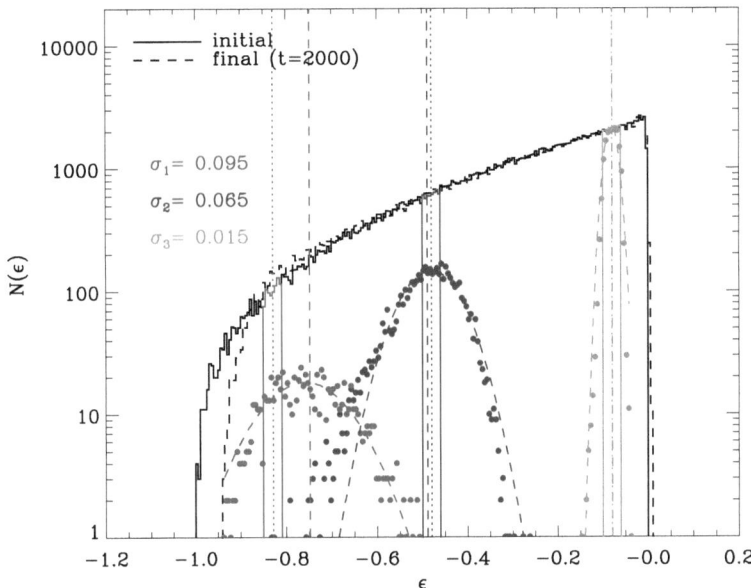

Figure 5.2: This panel shows the evolution of the energy distribution $N(\epsilon)$ of an one-component Hernquist sphere with $N = 160k$ particles and $G = M = a = 1$. After a time $t = 2000$ the most bound particles of the final profile (black dashed line) are 6% less strongly bound compared to the initial profile (solid black line). The three narrow histograms show the energy distribution for three dierent energy bins (red: $-0.85 < \epsilon < -0.80$, blue: $-0.50 < \epsilon < -0.45$, green: $-0.10 < \epsilon < -0.05$), which all become a gaussian distribution due to two-body encounters (cricles of the corresponding color). The gaussian fits (red, green and blue dashed lines) indicate, that the inner most particles (red) are more aected than the others, as the width σ of the fitted curves are higher for this energy bin. The vertical dashed and dotted lines indicate the initial and final mean binding energy of each bin, respectively.

to top, the number of particles gets doubled for each subsequent line and we can see, that the relaxation time at the spherical half-mass radius $r_{50,sph} = (1 + \sqrt{2})a$ (see Eq. 4.5) grows by a factor of ~ 13 if the particle number increases by a factor of 16. To highlight the effect of two-body relaxation we chose the model with $N = 160000$ particles (red line) and let it evolve for 2000 timesteps with a very small softening length of $= 0.01$.

Figure 5.2 shows the differential energy distribution of the initial galaxy model (solid black line) and the final one (dashed line) after $t = 2000$ timesteps. We can see, that the most bound particles at the left side of the distribution are finally less bound, because two-body encounters, which especially take place in the central, high density regions, lead to an equipartition of energies. Therefore, the most bound particles lose some of their energy to less bound particles. Furthermore, if the particles in a spheroid have different mass, the energy equipartition also leads to mass segregation, where the more massive particles tend to transfer energy to the less massive ones. Consequently the more massive particles sink towards the center and the lighter ones wander to larger radii. Next we look at the narrow bins for different binding energies (red/blue and green histograms in Fig. 5.2), which all evolve to Gaussians of different width . Again, the higher bound energy bins (red/blue histograms) get more broadened than the weakly bound bin (green histogram).

There are two main ways to reduce the effect of two-body relaxation. First, as already depicted in Fig. 5.1, an increasing number of particles significantly increases the relaxation time, and second, a larger softening length also limits the amount of scattering events. The drawback of the latter solution is, that one loses the information within two softening length, as the results in these regions are no longer reliable. However, in Section 4.3.3, we have already seen, that for galaxy models, consisting of unequal mass particles, it is crucial to adopt higher softenings to prevent mass segregation in the center. To quantify the effect of a larger softening length, we also evolved the same Hernquist sphere of Fig. 5.2 with $= 0.08$, which yields a much weaker broadening compared to the above simulation. The final width of the innermost bin is only $_1 = 0.057$ and therefore the effect of two-body relaxation is reduced by 40%.

Additionally, Fig. 5.3 depicts the depletion of the most central regions, which we could already see in Fig. 5.2. Due to equipartition of energy, caused by two-body relaxation, the central particles get slightly less bound and the innermost density profile becomes shallower.

5.2 Dynamical Friction & Tidal Stripping

Dynamical friction is a gravitational drag force, introduced by Chandrasekhar (1943). If a heavy point mass is traveling through a uniform background mass distribution, it attracts the surrounding particles and builds up an overdensity in its wake. This overdensity acts like a drag force onto the point mass, which consequently gets decelerated. On the other hand this implies a energy transfer from the satellite to the surrounding

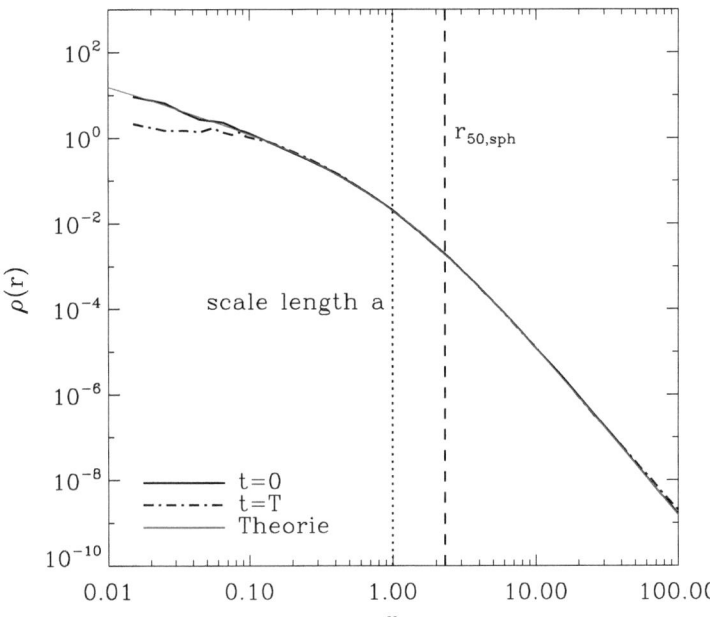

Figure 5.3: The black solid line and the black dashed-dotted line show the initial and final radial density distribution respectively. As the most bound particles go to higher energy the central density cusp within 10% of the scale length a (dotted line) gets depleted. The red solid line depicts the analytic profile of the Hernquist sphere (Eq. 4.16).

medium, as the energy of the combined system (satellite+background distribution) is conserved.

The original dynamical friction formula of Chandrasekhar (1943), describing the deceleration of infalling 'point masses' is,

$$\frac{d}{dt}\mathbf{v}_{orb} = -4\pi G^2 \ln(\Lambda) M_{sat} \rho_{host}(< v_{orb})\frac{\mathbf{v}_{orb}}{\mathbf{v}_{orb}}, \quad (5.3)$$

where Λ is the Coulomb logarithm (Chandrasekhar, 1943; Binney & Tremaine, 2008), $\rho_{host}(< \mathbf{v}_{orb})$ is the background density of all particles with velocities smaller than the orbital velocity $\mathbf{v}_{orb}$ of the satellite with mass M_{sat}. But this formula is based on three 'unrealistic' assumptions, that i) all particles and the satellite are point masses, ii) there is no self-gravity for the particles in the wake and iii) the background particle distribution is infinite, homogeneous and isotropic. However, by a more convenient choice of the Coulomb logarithm and restricting to minor mergers, where the satellite's mass is at maximum $\leq 20\%$ of the host galaxy, the dynamical friction force of Chandrasekhar (1943) is a viable approximation. Furthermore, the dynamical friction drag force F_{df} highly depends on the mass of the satellite M_s, as $F_{df} \propto M_s^2$. In numerical simulations, Boylan-Kolchin et al. (2008) has recently shown, that the mass loss of an infalling satellite is not negligible and has to be taken into account.

One way to unbind the particles of the satellite galaxy is violent relaxation, discussed in the next section, and *tidal stripping*. A simplified method to explain tidal stripping is the following. If a satellite galaxy with mass m is on a circular orbit around a massive point mass M with a distance R, it experiences an acceleration GM/R. But, as the satellite has a certain extension, the two boundaries at the farthest and the nearest end to the point mass M notice a different acceleration. If this tidal acceleration is higher than the binding energy of the lowest bound satellite particles, they can be stripped and leave the satellite's potential well (see also Mo et al. 2010). The radius, at which this tidal acceleration exceeds the binding energy of the particles is called *tidal radius* r_t. So far, due to many idealized assumptions, there are only very crude approximations to quantify r_t.

5.3 Violent relaxation

In contrast to two-body relaxation, violent relaxation is a physical mechanism, which was introduced by Lynden-Bell (1967). In a collisionless system, violent relaxation efficiently redistributes the energy of single stars due to local fluctuations of the gravitational potential,

$$\frac{d\mathrm{E}}{dt} = -\frac{d\Phi}{dt}, \quad (5.4)$$

where E is the energy per unit mass and $\Phi(x,y,z,t)$ is the gravitational potential of the whole system. This effect occurs on very short timescales, e.g. less than the free-fall time of the system (Bindoni & Secco, 2008).

5.3 VIOLENT RELAXATION

Although the theory of violent relaxation is not fully understood until now, we know that it plays a very important role during the coalescence of two or more galaxies. Therefore, we give a basic description of the original version of Lynden-Bell (1967) and show some more recent, slighlty dierent approaches. In Section 6.4.1 we try to figure out the eect of violent relaxation for numerical simulations.

5.3.1 Lynden-Bell's approach

The most basic quantity in stellar dynamics is the fine-grained distribution function (DF) or phase space density $f(\vec{x},\vec{v},t)$, which specifies the number of stars within an infinitesimal volume $d^3\vec{x}d^3\vec{v}$ at time t. Furthermore, it is convenient to introduce a coarse-grained distribution function F, which gives the average of the fine-grained DF in a small volume $\Delta^3\vec{x}\Delta^3\vec{v}$. Contrary to the fine-grained DF the coarse-grained one depends on the particular choice of partitioning the phase-space in which the volume elements $\Delta^3\vec{x}\Delta^3\vec{v}$ are defined. Additionally, only the evolution of the fine-grained DF can be described by the Boltzmann equation

$$\frac{df}{dt} = \frac{\partial f}{\partial t} + \vec{v}\frac{\partial f}{\partial \vec{x}} - \frac{\partial}{\partial \vec{x}}\frac{\partial f}{\partial \vec{v}} = 0. \tag{5.5}$$

Introducing a change in particle energy as described above, rearranges the orbits of stars and the system seeks for a new equilibrium configuration. Therefore we divide the 6 dimensional phase space in a big number of 'coarse-grained' macrocells n_i with equal volumes (Lynden-Bell, 1967). All these macrocells consist of a large number of even smaller microcells, where some of those are occupied by a phase element of particles. The latter volume is that fine, that it can adequately describe the fine-grained DF. Combining all macrocells results in a macrostate, which can be viewed as a discretized realization of the coarse-grained DF F of the system at time t. This means that $F(\vec{x},\vec{v},t)$, is defined as a discrete function on the ith macrocell

$$F_i(\vec{x},\vec{v},t) = \frac{1}{d^3\vec{x}d^3\vec{v}}\int_{d^3\vec{x}d^3\vec{v}} f d^3\vec{x}d^3\vec{v} = \frac{n_i n}{d^3\vec{x}d^3\vec{v}}, \tag{5.6}$$

where n is the number of particles in a phase element.

Now we calculate a functional $W\{n_i\}$, which gives all possible combinations of macrocells for a particular macrostate. Defining $S = \ln W$ as an Boltzmann entropy, the new statistical equilibrium state can be seen, as the macrostate, which maximizes the entropy S under the constraints of energy and mass conservation. Including the mass and energy constraints as Lagrange multipliers λ_1, λ_2, the maximization process can be written as,

$$\ln W - \lambda_1 N - \lambda_2 E = 0, \tag{5.7}$$

where $E = \sum_i n_i \epsilon_i$ is the total conserved energy (ϵ_i is the mean energy of all particles in the ith macrocell) and $N = \sum_i n_i$ is the total number of phase elements, which indicates mass conservation. Introducing $= \frac{n}{d^3\vec{x}\,d^3\vec{v}} = f(\vec{x},\vec{v},t)$ as the constant phase space density inside each phase element and applying Stirling's formula for big numbers Eq. 5.6 becomes:

$$F_i = \frac{n_i}{}\big|_{s=max} = \frac{}{\exp(_1 + _2\,_i)+1} \qquad (5.8)$$

If we consider $_2 \equiv\; \propto T^{-1}$ as an inverse temperature and $\mu = -\,_1/$ as a chemical potential, Eq. 5.8 yields

$$F_i = \frac{}{\exp[\;(_i-\mu)]+1}, \qquad (5.9)$$

which nearly resembles the Fermi-Dirac statistics of quantum mechanics.

After phase mixing (see next section 5.4), we are in the so-called *non-degenerate* limit, i.e. $F_i \ll\; = f(\vec{x},\vec{v},t)$, and Eq. 5.9 tends to a Maxwell-Boltzmann distribution,

$$F_i =\;\exp[-\;(_i-\mu)] = A\exp(-\,_i), \qquad (5.10)$$

where $A =\;\exp(\mu\;)$. This implies, that the final equilibrium state approaches an isothermal sphere. Unfortunately a physical system can never attain this state, as the isothermal sphere has infinite total mass. Therefore real systems undergo an *incomplete relaxation* process, which means, that violent relaxation stops very rapidly, as the potential fluctuations die out through e cient phase mixing.

5.3.2 Other approaches

Since the pioneering work of Lynden-Bell (1967), there a many other authors, who tried to improved the theory of violent relaxation. First of all, Shu (1978) argued, that the occupation of microcells by phase elements introduces unnecessary complications. He stresses the point, that stars can be seen as real particles in phase space and not as infinitesimal parts of a continuum. Therefore, he occupies the microcells by single star particles before he also maximizes the entropy to get the final equilibrium state. As consequence his solution also leads to a combination of Maxwell-Boltzmann distributions, but compared to Lynden-Bell, the velocity dispersions of each Maxwellian component does not depend on the inverse of the phase-space volume, but on the inverse of the particle mass.

On the other hand, both theories of Lynden-Bell and Shu imply mass segregation, as their macrocells have dierent masses. As consequence, the more massive elements migrate to the center, while the less massive once go to the outer parts of the system. To solve this problem, Kull et al. (1997) divides the phase space in macrocells with dierent volumes, but equal masses and finally the coarse-grained distribution function again is a combination of Maxwellians, but now, all cells are characterized by the same temperature.

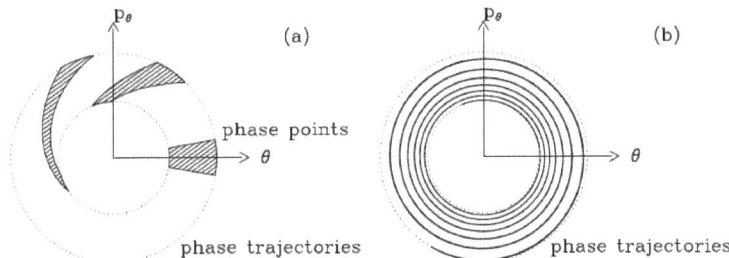

Figure 5.4: In the left panel, we can see the initial phase points at $p_i = 0$, where $f = F$. As the system evolves the phase points shear and the phase space volume gets thinner and occupies a larger regions. As more and more 'air' is mixed into the coarse-grained DF, it decreases with time. The right panel depicts a very late stage of the system, where the fine-grained DF f consists of infinitesimal thin lines. At this stage $F << f$ (see also Binney & Tremaine (2008)).

Apart from these examples, there a many other authors, who tried to get a description of the final state of a collisionless relaxing system by using entropy arguments (Nakamura, 2000; Stiavelli & Bertin, 1987). But recently Arad & Lynden-Bell (2005) argued, that all of them have limitations and they additionally show that the statistical-mechanical theories of violent relaxation are non-transitive. This non transitivity yields two dierent results, if a system either underg oes one violent relaxation process at once or two processes of comparable magnitude. Finally, Arad & Lynden-Bell (2005) conclude, that the already mentioned incompleteness of violent relaxation (see 5.3.1) is the most important reason for these shortcomings. One way to overcome the problems is to find a useful evolution equation for the coarse-grained DF.

Figure 5.4 shows a schematic realization of this scenario. The further evolution of the system can be described by the collisionless Boltzmann equation, which implies that the fine-grained DF f stays constant. Therefore the density of an infinitesimal volume around a phase points does not change.

5.4 Phase Mixing

In the previous section we have shown that phase mixing is responsible for the *incomplete violent relaxation* as it rapidly decreases the amplitudes of the potential fluctuations of, e.g. a merging event. Therefore the new equilibrium configuration after a merger never can reach the state of maximum entropy, which would be an isothermal sphere. On the other hand, it also increases the entropy of a system, as it decreases the coarse grained DF F.

The easiest way to illustrate the eect of phase mixing is by considering a system of

N pendulums, each of the same length L, though all of them have the same dynamical properties. At the beginning, all of them have are swung back by an angle $\varphi_{i=0,...,N}$, which all lie in a very small interval $\Delta \ll \varphi_i$. Now we define the fine-grained DF f for this system, which is initially the same as the coarse-grained DF F. If the pendulums are released they all have a dierent angular velocity $\dot{\varphi}_i$ with momenta $p_i = l\dot{\varphi}_i$, i.e. the pendulums with higher initial φ_i have lower momenta compared to those with smaller initial angles.

A macroscopic observer just can look at a cell of finite size and then he calculates the coarse-grained distribution in this cell. Initially $f = F$, but as the system evolves, the phase-space volume winds up in to infinitesimal thin filaments (see Figs. 5.4). Then the observer measures a much smaller phase-space density, because now, a lot of phase-space around the measured phase point is not occupied but empty. Finally the measured coarse-grained DF decreased a lot, compared to the initial value. This decrease of F, as the pendulums get out of phase is called *phase mixing*.

CHAPTER 6

RELAXATION AND STRIPPING: THE EVOLUTION OF SIZES AND DARK MATTER FRACTIONS IN MAJOR AND MINOR MERGERS OF ELLIPTICAL GALAXIES

In this chapter we investigate collisionless major and minor mergers of spheroidal galaxies in the context of recent observational insights on the structure of compact massive early-type galaxies at high redshift and their rapid size evolution on cosmological timescales. The simulations are performed as a series of mergers with mass-ratios of 1:1 and 1:10 for models representing pure bulges as well as bulges embedded in dark matter halos. For major and minor mergers, respectively, we identify and analyse two dierent processes, violent ralaxation and stripping, leading to size evolution and a change of the dark matter fraction. Violent relaxation - which is the dominant process for major mergers but not important for minor mergers - scatters relatively more dark matter particles than bulge particles to radii $r < r_e$. Stripping in minor mergers assembles satellite bulge particles at large radii in halo dominated regions of the massive host. This eect strong ly increases the size of the bulge into regions with higher dark matter fractions. For a mass increase of a factor of two, stripping in minor mergers increases the dark matter fraction within the eective radius by 75 per cent whereas relaxation in one equal-mass merger only leads to an increase of 25 percent. Compared to simple one-component virial estimates, the size evolution in minor mergers of bulges embedded in massive dark matter halos are very e cient. If such a two-component system grows by minor mergers only its size growth, $r \propto M$, will exceed the simple theoretical limit of $= 2$. Our results indicate that minor mergers of galaxies embedded in massive dark matter halos provide an interesting mechanism for a rapid size growth and the formation of massive elliptical systems with high dark matter fractions and radially biased velocity dispersions at large radii.

6.1 Introduction

Recent observations have revealed a population of very compact, massive ($\approx 10^{11} M_\odot$) and quiescent galaxies at z~2 with sizes of about $\approx$ 1kpc (Daddi et al., 2005; Trujillo et al., 2006; Longhetti et al., 2007; Toft et al., 2007; Zirm et al., 2007; Trujillo et al., 2007; Zirm et al., 2007; Buitrago et al., 2008; van Dokkum et al., 2008; Cimatti et al., 2008; Franx et al., 2008; Saracco et al., 2009; Damjanov et al., 2009; Bezanson et al., 2009). Elliptical galaxies of a similar mass today are larger by a factor of 4 - 5 (van der Wel et al., 2008) with at least an order of magnitude lower eective densities and significantly lower velocity dispersions than their high-redshift counterparts (van der Wel et al., 2005, 2008; Cappellari et al., 2009; Cenarro & Trujillo, 2009; van Dokkum et al., 2009; van de Sande et al., 2011). The measured small eective radii are most likely not caused by observational limitations, although the low density material in the outer parts of distant galaxies is dicult to detect (Hopkins et al. 2009a). Their clustering properties, number densities and core properties indicate that they are probably the progenitors of the most massive ellipticals and Brightest Cluster Galaxies (BCGs) today (Hopkins et al., 2009a; Bezanson et al., 2009).

Possible formation scenarios for such compact massive galaxies at redshifts $z \approx 2 - 3$ include gas-rich major disk mergers (Wuyts et al., 2010; Bournaud et al., 2011), accretion of satellites and gas, giant cold gas flows directly feeding the central galaxy in a cosmological setting (Kereš et al., 2005; Naab et al., 2007, 2009; Joung et al., 2009; Dekel et al., 2009; Kereš et al., 2009; Oser et al., 2010) or a combination of all of these. To explain the subsequent rapid size evolution dierent scenarios have been proposed (Fan et al., 2008; Hopkins et al., 2010; Fan et al., 2010). Frequent dissipationless minor and major mergers, which are also expected in a cosmological context, seem to be the most promising (Khochfar & Silk, 2006; De Lucia et al., 2006; Guo & White, 2008; Hopkins et al., 2010). Minor mergers, in particular, can reduce the eective stellar densities, mildly reduce the velocity dispersions, and rapidly increase the sizes, building up extended stellar envelopes (Naab et al., 2009; Bezanson et al., 2009; Hopkins et al., 2010; Oser et al., 2010, see however Nipoti et al., 2009a). Dissipationless major mergers will contribute to mass growth, however, their impact on the evolution of stellar densities, velocity dispersions and sizes is weaker (Boylan-Kolchin et al., 2005; Nipoti et al., 2009a). Observations and theoretical work also provide evidence that early-type galaxies undergo on average only one major merger since redshift ~ 2 (Bell et al., 2006b; Khochfar & Silk, 2006; Bell et al., 2006a; Genel et al., 2008) which would not be sucient to explain the observed evolution (Bezanson et al., 2009). In addition, major mergers are highly stochastic and some galaxies should have experienced no major merger at all, and would therefore still be compact today. However, such a population of galaxies has not been found yet (Trujillo et al., 2009; Taylor et al., 2010). Simulations in a fully cosmological context support the importance of numerous minor mergers for the assembly of massive galaxies. They might initially form at higher redshift during an early phase of in-situ star formation in the galaxy followed by a second phase dominated by stellar accretion (dominated by minor merging) onto the galaxy, driving the size evolution (Naab et al., 2009; Oser et al., 2010). Direct observational and circumstantial evidence has been recently presented in support of the minor merger scenario (van Dokkum et al., 2010; Trujillo et al., 2011).

Using the virial theorem, Naab et al. (2009) and Bezanson et al. (2009) presented a very simple way to estimate how sizes, densities and velocity dispersions of one-component collisionless systems evolve during mergers with different mass ratios. According to this simplified model assuming one-component systems on parabolic orbits, the accretion of loosely bound material (minor mergers) results in a significantly stronger size increase than predicted for major mergers (Naab et al., 2009). With the same approach Bezanson et al. (2009) found that eight successive mergers of mass ratio 1:10 can lead to a size increase of ~ 5, which corresponds to the observed difference between old compact galaxies and today's massive ellipticals. Of course, this is only valid for global system properties like the gravitational radii and total mean square speeds. The simple model is not including violent relaxation effects like mass loss, occurring during the encounter or non-homology effects which might change observable quantities.

Early papers on the interactions of spheroidal galaxies already discussed many of the above mentioned effects using N-body simulations of one-component spherical systems. White (1978, 1979), who carried out one of the first simulations of this kind, already found that relaxation effects are very efficient in equal-mass encounters and completely change the internal structure of the final remnants. First of all they contract in the central regions and build up diffuse envelopes of stars (see also Miller & Smith 1980; Villumsen 1983; Farouki et al. 1983), which leads to a break of homology. Furthermore, equal-mass mergers decrease population gradients due to the redistribution of particles in strong mixing processes (White, 1980; Villumsen, 1983), which breaks down in unequal-mass mergers, which even can enhance metallicity or color gradient (Villumsen, 1983). Farouki et al. (1983) also showed, that their multiple equal-mass mergers nicely recover the Faber-Jackson relation (Faber & Jackson 1976, see also Section 2.2) and that the velocity dispersion gets radially biased in the outer regions of the newly developed extended envelope. However, they all just used one-component models and therefore could not investigate the influence of the most massive part of a real galaxy, which is its dark matter halo. Naab & Trujillo (2006) and Hopkins et al. (2009b) already showed, that more realistic galaxy models, where the bulge is embedded in a dark matter halo, can change the size increase.

Although dissipationless minor mergers in general are able to increase sizes and decrease velocity dispersions, it is not clear if this scenario works quantitatively. Nipoti et al. (2003), who are among the first using spherical two-component models, argued that dry major and minor mergers alone cannot be the main mechanism for the evolution of elliptical galaxies, because their simulated merger remnants did not follow the Faber-Jackson (Faber & Jackson 1976) and Kormendy relations (Kormendy 1977), although they stayed on the fundamental plane. Nipoti et al. (2009a) found that dry major and minor mergers can bring compact early-type galaxies closer to the fundamental plane but the size increase was too weak for the assumed merger hierarchies.

Furthermore, dissipationless major merging introduces a strong scatter in the scaling relations, which are observationally very tight. Finally, Nipoti et al. (2009b) claim that early-type galaxies assemble only 50% of their mass via dry merging from $z \sim 2$ until now and the expected size growth of a factor of ~ 5 is hardly reproduced. However, especially in their minor merger sequences, they use very compact satellites, which might underpredict the eective size growth.

There are two main questions we address in this chapter. First, using highly resolved multiple equal-mass mergers, we investigate the impact of the massive dark matter halo on the dynamics of the final systems. Does it aect the central regions and can such mergers really change the central dark matter fraction, or is the increase just an artefact of the increasing radius (Nipoti et al., 2009b). Second, we revisit, whether dissipationless minor mergers are really too weak to fully account for the observed evolution of compact early-type galaxies (Nipoti et al., 2003, 2009a) and the implied size growth. Using more realistic two-component models, we are able to draw conclusions about the change of internal structure for the galaxies in both merging scenarios.

This chapter is organized as follows. First, in section 6.2 we give an overview of our initial galaxy setup and the employed numerical methods, before we highlight the virial predictions in section 6.3. In Section 6.4 and 6.5 we show the results for major and minor mergers, respectively. Finally, we summarize and discuss our findings in section 6.6.

6.2 Numerical Methods

6.2.1 Galaxy Models

For the initial galaxy models we assume spherical symmetric, isotropic Hernquist density profiles (Hernquist 1990) for both, the luminous and the dark matter (bulge+halo) component,

$$_i(r) = \frac{M_i}{2} \frac{a_i}{r(r+a_i)^3}, \qquad (6.1)$$

where $_i, M_i$ and a_i are the density, the mass and the scale length of the respective component i. The potential is

$$_i(r) = -\frac{GM_i}{r+a_i}, \qquad (6.2)$$

with the gravitational constant G.

On the one hand the projected Hernquist profile is a reasonable approximation of the $R^{1/4}$ law (de Vaucouleurs 1948) for the luminous component (its Sersic index however is closer to $n \sim 2.6$, see Naab & Trujillo 2006). On the other hand it is a good representation of the Navarro et al. (1997) profile for the dark matter component. Therefore we consider the Hernquist density distribution a su ciently realistic description for the luminous and dark matter distributions of a typical elliptical galaxy.

For simplicity, we assume isotropy of the velocity distribution to construct a stable initial configuration, i.e. the bulge and halo component are in dynamical equilibrium. We compute the distribution function (DF) f_i for each component i, using Eddington's formula (Binney & Tremaine, 2008),

$$f_i(E) = \frac{1}{\sqrt{8}\pi^2} \int_{\Psi=0}^{\Psi=E} \frac{d^2\rho_i}{d\Psi_T^2} \frac{d\Psi_T}{\sqrt{E-\Psi_T}}, \qquad (6.3)$$

where ρ_i is the density profile of component i, E is the relative (positive) energy and Ψ_T is the total gravitational potential $\Psi_T = \Psi_*(+\Psi_{dm})$. Solving distribution functions is in general more complicated than using Jeans equations, but results in more stable initial conditions (Kazantzidis et al., 2004). Only for a few models, e.g. the one-component Hernquist sphere (Hernquist, 1990), the distribution function can be computed analytically. For a two-component model (bulge+halo) we have to calculate f_i numerically. As there is no analytic expression for $\rho_i(\Psi_T)$ (see Eq. 6.3) we have to transform the integrand of Eq. 6.3 to be a function of radius r,

$$\frac{d^2\rho_i}{d\Psi_T^2} d\Psi_T = \left(\frac{d\Psi_T}{dr}\right)^{-2} \left[\frac{d^2\rho_i}{dr^2} - \right.$$
$$\left. - \left(\frac{d\Psi_T}{dr}\right)^{-1} \frac{d^2\Psi_T}{dr^2} \frac{d\rho_i}{dr}\right] \frac{d\Psi_T}{dr} dr. \qquad (6.4)$$

This procedure always results in an analytical expression for the integrand, even for more general γ-profiles (Dehnen, 1993) with different slopes for the density distribution. As a consequence, the integration limits also have to change, e.g. $\Psi(r) = 0$ becomes $r = \infty$ and $\Psi(r) = E$ has to be solved (numerically) for the radius r.

Once we have computed the DF we can randomly sample particles with radii smaller than a given cut-off radius and random velocities, which are smaller than the maximum escape velocity. Then the particle configuration for the galaxy is established using the Neumann rejection method (see also Chapter 4).

The one component model is described by two parameters, the scale length a_* and the total mass M_*. For the two component models, including dark matter, we additionally introduce the dimensionless parameters μ and β for the scale length of the halo $a_{dm} = \beta a_*$ and its mass $M_{dm} = \mu M_*$.

6.2.2 Model Parameters and Merger Orbits

For the total dark matter to stellar mass ratio we assume $\mu = M_{dm}/M_* = 10$ and the ratio of the scale radii is $\beta = a_{dm}/a_* = 11$, for all simulations with two-component models. We perform a set of simulations for two different scenarios with $M_* = a_* = 1.0$. In the major merger scenario we simulate equal-mass mergers of initially identical, spherically symmetric one- or two-component models on zero energy orbits. The encounters have parabolic orbits with and without angular momentum (head-on). For higher merger generations we duplicate the merger remnant after reaching dynamical

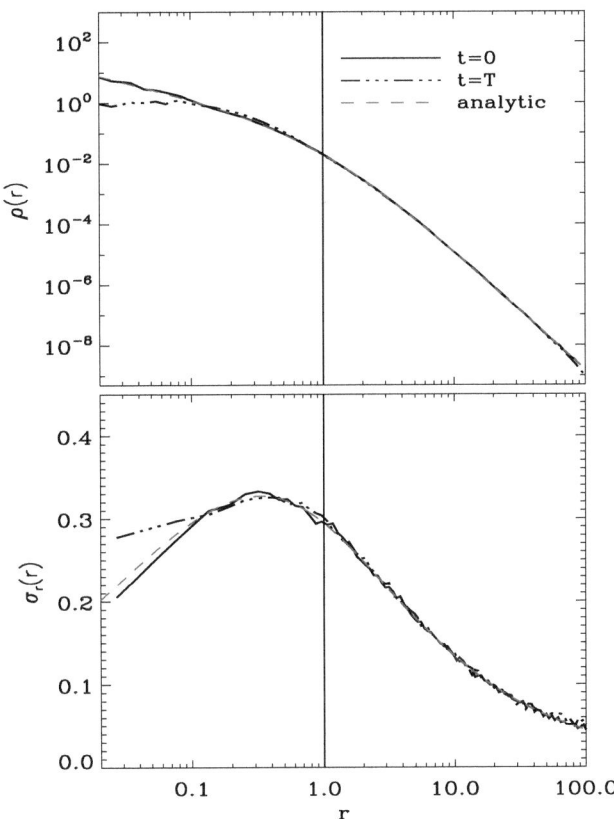

Figure 6.1: Top panel: Radial density profile for a Hernquist distribution (160k particles). The black solid line illustrates the initial profile (t=0) and the dashed-dotted line the final profile ($t = T \sim 100 \times t_{dyn}$, where t_{dyn} is the dynamical time at the spherical half-mass radius). The analytical profile is indicated by the red dashed line. Inside 10% of the scale length (vertical solid line) the system is aected (increase in dispersion and decrease in density) by relaxation eects. However, over all, the system is dynamically stable for at least 100 dynamical times. Bottom panel: Initial (solid line) and final (black dashed line) radial velocity dispersion profile.

equilibrium at the center and merge them again on orbits with the same energy but dierent infall directions. In total, we simul ate three generations of head-on equal-mass mergers, and two generations with angular momentum (see also table 6.1).

In the minor merger scenario our simulation sequences start with an initial mass ratio of 1:10, i.e. $M_{host} = 10 M_{sat}$ and a stellar scale radius of the satellite of $a_{*,sat} = 1.0$. On the first glance this choice for the satellite's scale radius seems very unrealistic and does not agree with any observed mass-size relation (see Fig. 2.1), but recent observations show that the sizes of less massive ellipticals converge at an eective radius of $r_e \sim 1 kpc$ (Misgeld & Hilker, 2011). Therefore the satellite galaxies have the same size as the compact early-type hosts, although they are an order of magnitude less massive. For comparison, we made two sequences (with one- and two-component models), of head-on minor mergers, where the satellite's scale radii are half the host's scale radius $a_{*,sat} = 0.5$, though the satellites lie on an extrapolation of the observed mass-size relation of Williams et al. (2010) at $z = 2$ (see Fig. 2.1 and table 6.1). The host galaxy for the next generation is the virialized end product of the previous accretion event. This host is merged with a satellite identical to the first generation. The mass ratio for this merger is now 1:11. We repeat this procedure until the host galaxy has doubled its mass, i.e. 10 minor mergers. The final mass ratio of the merger is 1:19. Again we simulate one- and two-component mergers with zero (head-on) and non-zero angular momentum. As the mergers of the bulge+halo model with angular momentum are computationally expensive we only simulate 6 generations.

For all head-on mergers we separate the centers by a distance d and assign them a relative velocity $v_{rel} = 2\sqrt{GM_h/d}$, where M_h is the total attracting mass of the host galaxy within radius d. This velocity corresponds to an orbit with zero energy and zero angular momentum, i.e. the galaxies will have a zero relative velocity at infinite distance. The distance d is always large enough to obtain virialized remnants at the end of each generation. As the merger remnants after the first generation will not be spherical anymore, their mutual orientation is randomly assigned at the beginning of each new merger event.

For the mergers with angular momentum we set the impact parameters to half of the spherical half-mass radius of the host's bulge and separate the galaxies far enough so that the initial overlap is very small.

6.2.3 Simulations and Stability Tests

All simulations were performed with VINE (Wetzstein et al., 2009; Nelson et al., 2009), an e cient parallelized tree-code. We use a spline softening kernel with a softening length = 0.02 for all runs. In general, the softening length depends on the particle number (e.g. Merritt 1996; Dehnen 2001) and we found = 0.02 to be the best value looking at the balance between computational costs and stability of the models. For the major merger simulations the seed galaxy consists of $N_* = 1.6 \times 10^5$ bulge particles for the one-component (bulge only) model and $N_* = 2 \times 10^4$ for the two-component model, which has an additional halo of $N_{DM} = 2 \times 10^5$ particles of the same mass. For

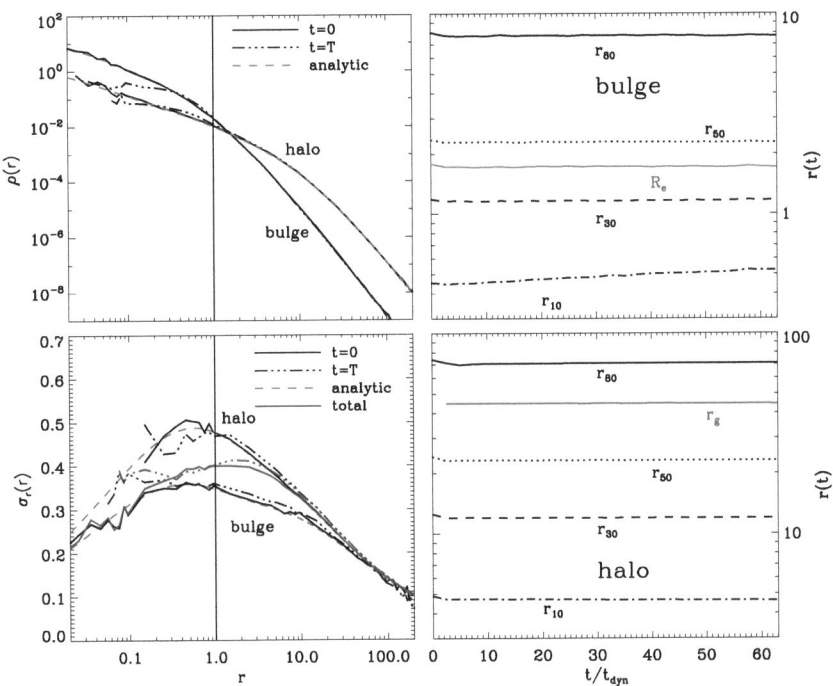

Figure 6.2: Top left panel: Radial density profile for the two-component realization with 1100k particles. The bulge to halo mass ratio is 1:10. The solid black lines illustrate the initial profiles (t=0) of the bulge and the halo and the dashed-dotted lines their final profiles ($t > 60 t_{dyn}$). The analytic Hernquist profiles are indicated by the red dashed line. Bottom left panel: Radial velocity dispersion of the total system (blue solid: initially, blue dashed-dotted: final), the bulge and the halo separately. Inside $0.3 a_*$ the model is aected by two-body relaxation, but overall it is stable. Right panels: Time evolution of the radii enclosing 80%, 50%, 30% and 10% mass (black lines from top to bottom) of the bulge (top panel) and halo (bottom panel). The red lines show the eective radius of the bulge (upper panel) and the gravitational radius of the whole system (bottom panel). Except the 10% radius of the bulge, which shows a slight increase, all mass radii stay constant over > 60 dynamical times.

Run	Gen.	$a_{*,sat}$	$M_{ub}(\%)$	$M_{*,ub}(\%)$
B1ho	3	1.0	12.3	12.3
B1am	2	1.0	15.0	15.0
HB1ho	3	1.0	10.1	2.5
HB1am	2	1.0	8.8	2.0
B10hod	10	1.0	21.8	21.8
B10amd	10	1.0	20.9	20.9
B10hoc	10	0.5	21.7	21.7
HB10hod	10	1.0	35.6	20.4
HB10amd	6	1.0	19.5	7.9
HB10hoc	10	0.5	20.9	7.2

Table 6.1: This table gives the name of the hierarchy (1st column), the number of generations (2nd), the initial scale radius of the satellite (3rd), the amount of unbound mass of the total final remnant (4th) and the corresponding stellar mass loss (5th). The name can be explained as followed; B/HB: bulge or halo+bulge, 1/10: major/minor merger, am/ho: orbit with/without angular momentum. In the case of the minor merger scenarios, c/d indicates wether we chose a compact or diuse satellite.

the accretion scenario, the one- and two-component host galaxies both have $N_* = 10^5$ bulge particles and the latter has $N_{DM} = 10^6$ halo particles. The satellites have ten times less particles for all components.

In Figure 6.1 we demonstrate the stability of the bulge only model with 160k particles by comparing the initial and final (100 dynamical times) density (top panel) and radial velocity dispersion (bottom panel) as a function of radius to the analytical solution of the Hernquist sphere. In general the model is very stable after the whole simulation time, except in the innermost parts, where two-body relaxation becomes important. At the highest densities, inside 10% of the scale radius the relaxation time t_{relax} of the model is very small. Consequently two-body encounters change the central particle's energy and deplete the high density regions. Looking at the initial and final number of particles within $0.1a_*$, we find that half of the particles escape this region and go to lower binding energies, which is in good agreement to the results of Section 4.3. However, at larger radii the models are very stable with a very good agreement with the analytical solution. The radii enclosing 30, 50 and 80 per cent of the stellar mass stay perfectly constant.

The stability of the two-component system is demonstrated in Fig. 6.2. Our initial model, constructed of two Hernquist spheres, is again very stable over a long simulation period of more than 60 dynamical times. The density and the velocity dispersion do not change significantly. Again the innermost regions of the bulge distribution are aected by two-body relaxation as the particle number is similar to the former one-component sphere. However, looking at the mass radii of the bulge and the halo we observe no significant changes. The apparent contraction of the mass radii enclosing 80% or 50%

is less than 5% for the halo and less than 2% for the bulge and is an artefact of the cut-o radius (see Section 4.3). The gravitational radius and the eective radius which we use in this paper stay constant.

Therefore we conclude that our results are not aected by two-body relaxation and other numerical artifacts on the initial conditions.

6.3 Analytic Predictions

Naab et al. (2009) found a very simple prescription of how stellar systems evolve during a merger event. This will be extended later on and therefore is reviewed briefly. Using the virial theorem and assuming energy conservation we can approximate the ratios of the initial to the final mean square speed $\langle v_{i/f}^2 \rangle$, gravitational radius $r_{g,i/f}$ and density $_{i/f}$ of a merging system. According to Binney & Tremaine (2008) the total energy of a system is

$$\begin{aligned} E_i &= K_i + W_i = -K_i = \frac{1}{2} W_i \\ &= -\frac{1}{2} M_i \langle v_i^2 \rangle = -\frac{1}{2} \frac{GM_i^2}{r_{g,i}}, \end{aligned} \quad (6.5)$$

where E_i and M_i are the system's initial total energy and mass. The gravitational radius is defined as

$$r_{g,i} \equiv \frac{GM_i^2}{W_i}, \quad (6.6)$$

with the total potential energy W_i. Now we define E_a, M_a, $r_{g,a}$ and $\langle v_a^2 \rangle$ as the energy, mass, gravitational radius and mean square speed of the accreted system. Furthermore, $= M_a/M_i$ and $= \langle v_a^2 \rangle / \langle v_i^2 \rangle$ are the dimensionless mass and velocity fractions respectively. By combining these assumptions with equation 6.5 we obtain

$$\frac{\langle v_f^2 \rangle}{\langle v_i^2 \rangle} = \frac{(1+\ )}{(1+\ )}, \quad (6.7)$$

$$\frac{r_{g,f}}{r_{g,i}} = \frac{(1+\ )^2}{(1+\ )}, \quad (6.8)$$

$$\frac{_f}{_i} = \frac{(1+\ )^3}{(1+\ )^5}, \quad (6.9)$$

for the ratios of the final to initial mean square speed, gravitational radius and density. In the very simple case of an equal mass merger of two identical systems $=\ =1$, r_g is doubled (Eq. 6.8), $\langle v^2 \rangle$ stays constant (Eq. 6.7) and decreases by a factor of 4 (Eq. 6.9). If we use equations 6.7-6.9 for a minor merger scenario, where $\langle v_a^2 \rangle << \langle v_i^2 \rangle$ and $<< 1$, the size of the final system can increase by a factor of ~ 4, as $r \propto M^2$.

Additionally, the final velocity dispersion and density are reduced by a factor of 2 and 32, respectively. These changes are quantitatively in good agreement with observations.

However, this simple, analytic model suers from a number of limitations, apart from the restrictions to parabolic orbits and collisionless systems. The eect of violent relaxation (Lynden-Bell, 1967) in the rapidly changing potential during the merger will scatter particles in the energy space, making some more bound and unbind others from the system. Thus, energy is not perfectly conserved. Additionally, realistic spheroidal galaxies are composed of two collisionless components, dark and luminous matter with dierent spatial distributions, which are expected to react dierently to a merger event. In the following we investigate the eect of violent relaxation and dark matter for spheroidal mergers with mass ratios of 1:1 and 1:10.

6.4 Major Mergers

In the case of equal-mass mergers, the eect of violent relaxation is very strong and has a significant eect on the dierential energy distributions of the remnants. The left panel of Fig. 6.3 indicates, that the initial narrow distribution (black line) becomes broader with each generation in a way that tightly bound particles get even more bound and some weakly bound ones gain enough energy to escape the galactic potential. The theoretical framework of violent relaxation is very complex and since the pioneering work of Lynden-Bell (1967), there have been many approaches to develop a more viable theory, which can describe the final equilibrium configuration of a violently relaxing system (e.g. Shu 1978; Nakamura 2000). In the following, we briefly repeat the original approach of Lynden-Bell (1967) (Section 5.3.1), before we discuss our results with respect to another slightly dierent approach of Spergel & Hernquist (1992).

6.4.1 Violent relaxation

During the approach of two collisionless systems the total gravitational potential varies with time, which leads to a non-conservation of energy of single particles, (Lynden-Bell, 1967; Spergel & Hernquist, 1992)

$$\frac{d\epsilon}{dt} = -\frac{\partial \phi}{\partial t}, \quad (6.10)$$

where ϵ is the energy per unit mass. In accordance with the time dependent virial theorem,

$$\frac{1}{2}\frac{d^2 I}{dt^2} = 2T + V, \quad (6.11)$$

where I is the moment of inertia tensor, the galaxy will convert its total potential energy V into kinetic energy T and back. In equilibrium, $\ddot{I} = 0$ so $T = -E, V = 2E$, with $E = T + V$ being the total energy. Away from equilibrium the total energy E is

6.4 Major Mergers

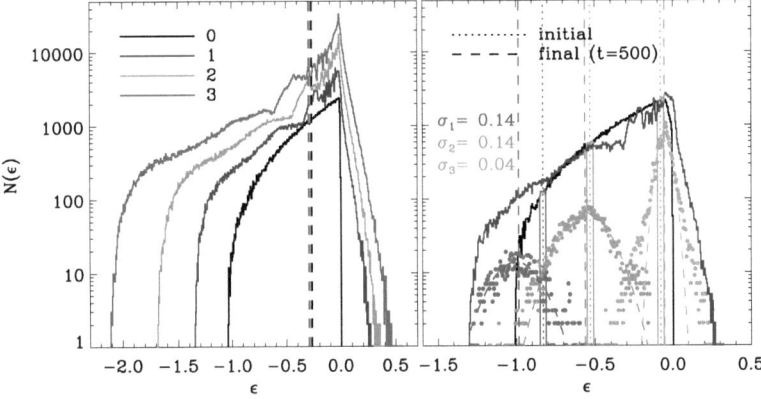

Figure 6.3: Left panel: Energy distribution of the initial host galaxy and three generations of head-on major mergers (B1ho). The initial distribution (solid black line) is broadened by each merger towards lower and higher (escapers) energies. Finally the most bound particles have two times their initial binding energy. The mean energy of the total system stays constant (vertical dashed lines). Right panel: Here we investigate the eect of violent relaxation on the particles of one galaxy. The overall evolution is the same as in the left panel, i.e. the final width is the same. Looking at three dierent energy bins at high (red, $-0.85 < \epsilon < -0.81$), intermediate (green, $-0.55 < \epsilon < -0.51$) and low binding energies (light blue, $-0.1 < \epsilon < -0.06$), they show a dierent evolution. As the innermost and the intermediate bin suer more from violent relaxation, their final width is much broader compared to the outermost bin. Additionally, due to the escaping particles, the latter bin cannot be fitted by a gaussian, whereas the other two bins can be fitted by a gaussian with a width of $\sigma = 0.14$. The vertical lines indicate, that the mean of the most bound particle is shifted to even higher binding energies (to the left) compared to the other two bins.

constant, but T and V scatter about these values, which widens the dierential energy distribution $N(E)$, where $N(E)$ gives the number N of stars within an energy interval $E + dE$.

This evolution is illustrated in the left panel of Fig. 6.3, where the energy distribution $N(\ )$ becomes wider with each higher merger generation. Spergel & Hernquist (1992) did similar simulations and find exactly the same results. Using the Ansatz that violent relaxation can be approximated by scattering eects of single particles, they find an analytic prediction of the final equilibrium configuration of an equal-mass merger. Furthermore, they illustrate, that the probability function of the scattering eects becomes gaussian. In the right panel of Fig. 6.3 we select three dierent particle bins, with low ($-0.06 < E < -0.1$), intermediate ($-0.55 < E < -0.51$) and high ($-0.85 < E < -0.81$) binding energies from the initial host galaxy. After the final merger, these narrow bins are broadened significantly and the two innermost bins can be perfectly fitted by a gaussian of width $= 0.14$. Due to escaping particles, the outermost bin does not develop a gaussian shape. Furthermore, the vertical lines indicate, that the mean energy of the innermost bin (red lines) is significantly shifted to higher binding energies, whereas the mean energies of the intermediate and weakly bound particles only show a small shift to higher and lower binding energies, respectively. Further investigation of the innermost energy bin indicates that, at the first close encounter, it is shifted to higher binding energies (left panel Fig. 6.4, $t = 110 \rightarrow 120$) and broadened by a factor of 3 ($= 0.2 \rightarrow 0.6$). This can be explained by a sudden deepening of the potential, as each galaxy experiences a doubled mass in its center during their closest approach (see also right panel of Fig. 6.4). Afterwards, the two galaxies separate and the mean energy goes nearly back to its original value, without further broadening ($t = 130$). During the second close encounter, this scenario is repeated, i.e. the highest energy bin is shifted to even higher binding energies accompanied by a strong widening ($t = 130 \rightarrow 140$), before it oscillates back into a less bound state ($t = 140 \rightarrow 150$). But now, the particle bin resides at a slightly higher mean binding energy and does not go back to its initial position. In the right panel of Fig. 6.4 we depict the evolution of the mean potentials of the three energy bins shown in Fig. 6.3, which oscillate strongly for the strong bound particles (red, green line). Additionally, for the innermost bin, the energy shifts and broadening is obviously correlated to these potential fluctuations. On the other hand, we can see, that these fluctuations vanish rapidly, due to phase mixing, which results in a so called incomplete relaxation (Lynden-Bell, 1967; Shu, 1978).

Checking the eect of two-body relaxation for the isolated host during the same time interval yields a much weaker eect (e.g. $= 0.02 \rightarrow 0.03$ between $t = 110 \rightarrow 160$). To get a comparable broadening by two-body relaxation, we have to run the simulation for more than 2000 time steps (see Section 5.1)

Therefore, we conclude, that violent relaxation yields a significant widening of single energy bins, which results in a much broader dierential energy distribution. As a consequence, some of the weakly bound particles acquire positive binding energies and can escape the remnants potential. On the other hand, violent relaxation oers new energy states by deepening the total gravitational potential, into which the most bound

6.4 Major Mergers

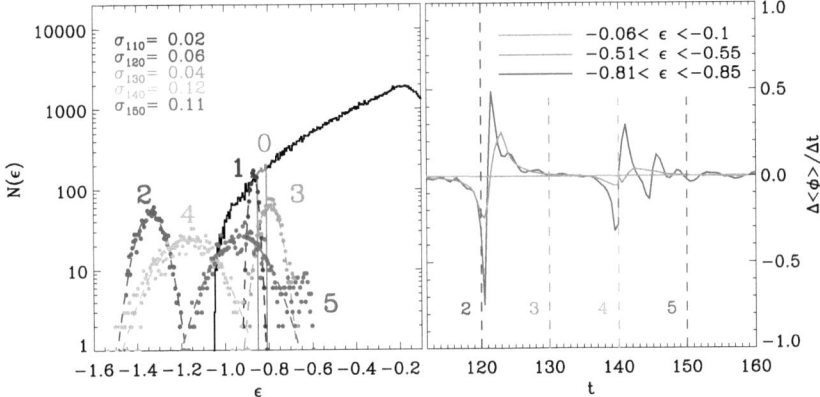

Figure 6.4: Left panel: Time evolution for the energy distribution of the most bound energy bin in Fig. 6.3 (red bin). The initial distribution at $t = 100$ (0) is slightly broadened to $\sigma_{110} = 0.02$ (1) by two-body relaxation. At the first close encounter $t \approx 120$ (2) the potential changes rapidly (see also right panel), the particles get shifted to higher binding energies and the energy distribution widens to $\sigma_{120} = 0.06$. As the two galaxies fly away from each other, the potential increases and the particles go back to lower binding energies (3) without further broadening of the distribution. During the second close encounter, this scenario repeats, i.e. the distribution becomes broader when the mean energy is shifted to a higher binding energy (4). After $t = 150$ (5) the central regions show only negligible potential fluctuations and the particle distribution is slowly affected by two-body relaxation. In the merger, the innermost energy bin is broadened from $0.02 - 0.1$, while in isolation, the same bin only is slightly affected by two-body relaxation ($\sigma = 0.02 \rightarrow 0.03$) in the same time interval. Right panel: The fluctuations of the mean potential of the three energy bins in Fig. 6.3 are strongest in the central regions (red bin) and decrease for lower bound particles (green/light blue bin, Fig. 6.3). For the outermost bin, the fluctuations are too small to be visible in this panel.

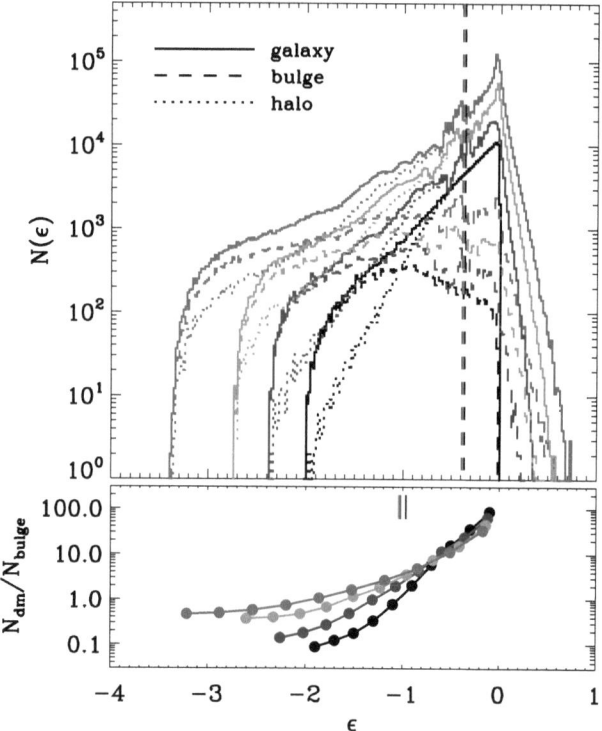

Figure 6.5: Top panel: Dierential energy distribution for two-component initial conditions (black) and after the first (blue), second (green), and third (red) generation. The solid lines depict the distribution of all particles, the dashed lines of the bulge and the dotted lines of the halo. In general, the evolution of the total system is similar to the one-component system (left panel, Fig. 6.3). The mean binding energies of the remnants (vertical dashed lines) stay constant. However, relatively more dark matter particles than bulge particles are scattered to low energies (more bound), thereby increasing the central dark matter fraction. Bottom panel: The central ratio of dark matter to bulge particles, calculated for ten energy bins, increases with each generation. Finally, there are nearly as many dark matter as bulge particles in the innermost bin. The small vertical lines at the top of the panel indicate the energy of the 50% most bound bulge particles, which stays constant after the first merger.

particles are scattered. As result, they become even more tightly bound.

In the top panel of Fig. 6.5 we can see, that the overall energy distribution of the same merger history of two-component models (solid lines) evolves the same as for the one component models, i.e. the tightly bound particles go to states with even higher binding energies and some low bound ones get unbound (positive energies). But looking at the dierent components separate ly, we can see, that with each generation, the number of dark matter particles in the highly bound regions increase more than the number of bulge particles. This behavior can also be seen in the bottom panel of Fig. 6.5, as the fraction of dark matter to bulge particles converges to one in the central regions. As the energy of the 50% most bound bulge particles, indicated by the small vertical lines at the top of this panel, only changes for the first generation, this indicates that the structure of the system changes, which implies a 'real' change of the dark matter fraction. The fact, that more dark matter than bulge particles wander to higher binding energies is illustrated in Fig. 6.6, where we take a closer look at the remnant of the first merger generation. Overall, the amount of dark matter, going to higher binding energies, is significantly larger for most of the ten energy bins, especially for ≥ -1.2, which is the region where the initial number of halo particles equals the number of bulge particles (see top panel Fig. 6.5). Although there are only very few dark matter particles in the initially most bound regions, a non-negligible amount occupies the finally most bound state of the remnant.

Finally, we can say, that violent relaxation rearranges the distributions of dark and luminous matter in energy space, which yields a higher dark matter fraction in the center of the final system, which is not just an eect of the increasing mass radius.

6.4.2 Velocity dispersion

One crucial condition of violent relaxation is, that each merging system evolves to a state of higher entropy. Ideally, if violent relaxation would be complete, the final state of a relaxing system should be the maximum entropy state, the isothermal sphere. In reality, phase mixing damps the potential fluctuations of a merger rapidly, and violent relaxation is incomplete, which results in a final equilibrium distribution which does not reach a maximum entropy state.

Nevertheless the left panel of Fig. 6.7 shows that the radial velocity dispersion of the final remnant (red line) in the one-component merger scenario can well be fitted by a Jae profile (Jae , 1983), which resembles the inner parts of the singular isothermal sphere (Tremaine et al., 1994). In the case of two-component models (right panel Fig. 6.7) we get the same result, i.e. the total (solid line) and also the bulge profile (dotted line) get close to a Jae profile. Spergel & Hernquist (1992) find the same trend for only one generation of a head-on equal mass merger, which leads to the conclusion, that although the maximum entropy state can never be reached, each major merger event brings the final system closer to this state. Another proof, that the structure of the systems changes is shown by the structure parameter c in both panels of Fig. 6.7. As the parameter changes and the final energy distribution of both merger hierarchies

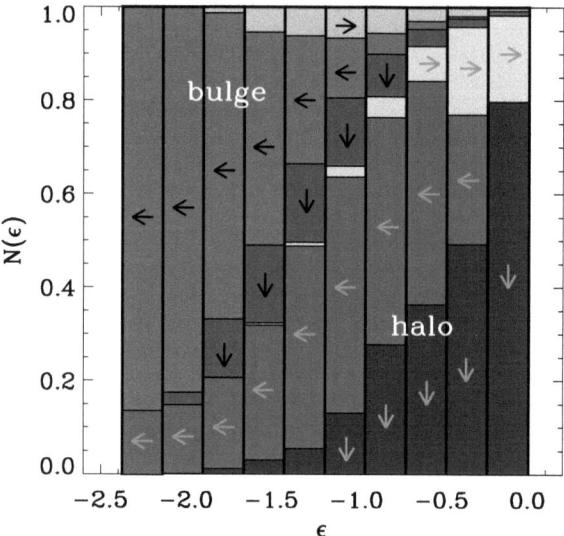

Figure 6.6: This panel shows, in which bins bulge and halo particles get scattered during the first merger (black to blue line in Fig. 6.5). Initially, the binding energies < -2.0 are not occupied, which means that the bin left to this energy consists of particles, which come from higher energies, highlighted by a the intermediate blue (halo) and red (bulge) histograms with left pointing arrow. At energies ≥ -2.0 there are also particles which stay in its initial energy bin, which is depicted by the dark blue (halo) and dark red (bulge) histograms with arrows pointing downwards. The light blue/red regions with the right pointing arrows at the top right of the figure show all the halo/bulge particles which go from higher to lower binding energies. All energy bins are normalized to each bins total number of particles. Overall we can see, that in most of the final energy bins (especially at ≥ -1.2), much more dark matter than bulge particles come from higher binding energies, which is consistent with the increasing fraction of central dark matter particles (bottom panel Fig. 6.5).

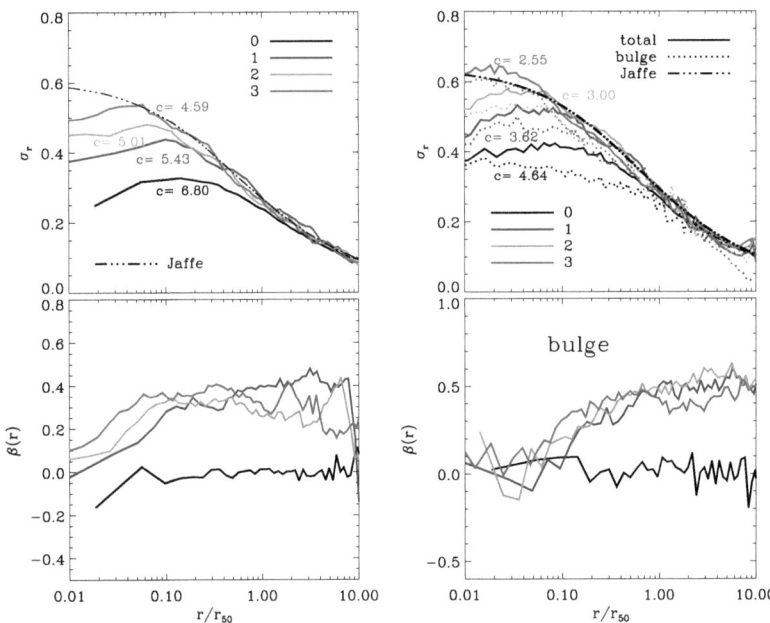

Figure 6.7: Top left: Final radial velocity dispersions of the one-component merger generations (B1ho). The colors depict the dierent generations. The dashed-dotted line indicates the velocity dispersion of a Jae profile, which has the same scale length and mass as the last remnant. The latter profile resembles the inner parts of the singular isothermal sphere which is a very good fit to our last merger remnant. As the structure parameters c decreases with each generation homology is not preserved. Bottom left: Anisotropy parameter (eq. 6.12) against radius of three generations of one-component equal-mass mergers. As > 0 for higher generations the remnants become radially anisotropic over the whole radial range. The half-mass radius r_{50} is the radius of the sphere, which includes half of the bound system mass. Top right panel: Here we show the same as in the left panel for the head-on two-component mergers (HB1ho). The bottom right panel depicts the anisotropy parameter for the bulge. Here, r_{50} is the spherical half mass radii of the total (top) and stellar (bottom) bound remnants.

have a dierent shape (see above), homology is not preserved.

Furthermore, in the bottom panels of Fig. 6.7 we can see, that the initially isotropic remnants become radially anisotropic over nearly the whole radial range. Already after the first merger generation the anisotropy parameter (Binney & Tremaine, 2008)

$$\beta = 1 - \frac{\sigma_\theta^2 + \sigma_\phi^2}{2\sigma_r^2}. \tag{6.12}$$

gets positive, which indicates radial anisotropy. But this result is not surprising, as we only use orbits with very small or zero impact parameter. Consequently, most of the material falls in radially, which then causes the velocity distribution to become radially biased (see also Boylan-Kolchin & Ma 2004).

6.4.3 System Evolution

In Fig. 6.8 we compare the simple theoretical predictions for the gravitational radius (Eq. 6.8), the density (Eq. 6.9) and the mean square speeds (Eq. 6.7) to the dierent major merger scenarios.

The top panel shows the evolution of the mean square speeds of the one- and two-component equal mass mergers with the total bound mass of the system. According to Eq. 6.7 the mean square speed should remain unchanged (dashed line), but obviously it increases with each generation. As a consequence, the growth of the total gravitational radius (middle panel of Fig. 6.9) and the density decrease (bottom panel of Fig. 6.9) is weaker than expected. The same trend was reported by Nipoti et al. (2003, 2009a) who argued that the simple analytical prediction is only valid for an idealized case without escaping particles. In the 4th column of table 6.1 we can see, that the amount of unbound mass after the merger is not negligible and adds up to about 12, 15, 10 and 9% per cent of the total mass for the scenarios B1ho, B1am, HB1ho and HB1am respectively. As the coalescence time for the one-component mergers with angular momentum (B1am) is longer than for the head-on orbits, this hierarchy suers most from violent relaxation and has the largest amount of escaping particles. The same eect can be seen for the two-component case, where two merger generations with angular momentum have nearly as much mass loss as three generations of the head-on counterparts ($\approx 9\%$ compared to $\approx 10\%$). Looking at the last column of table 6.1 we can see, that nearly all escaping particles for the bulge+halo models are from the halo, as nearly no stellar mass gets lost ($M_{*,ub} < 3\%$).

Going back to the analytic predictions and taking the eect of escapers into account (see also Nipoti et al., 2003) we can re-write the energy equation using the energy of the bound final system E_f and the energy of the escaping particles E_{esc},

$$E_f + E_{esc} = E_i + E_a. \tag{6.13}$$

We assume that the escaping particles have essentially zero potential energy, so that

$$E_{esc} = +\frac{1}{2}M_{esc}\langle v_{esc}^2\rangle. \tag{6.14}$$

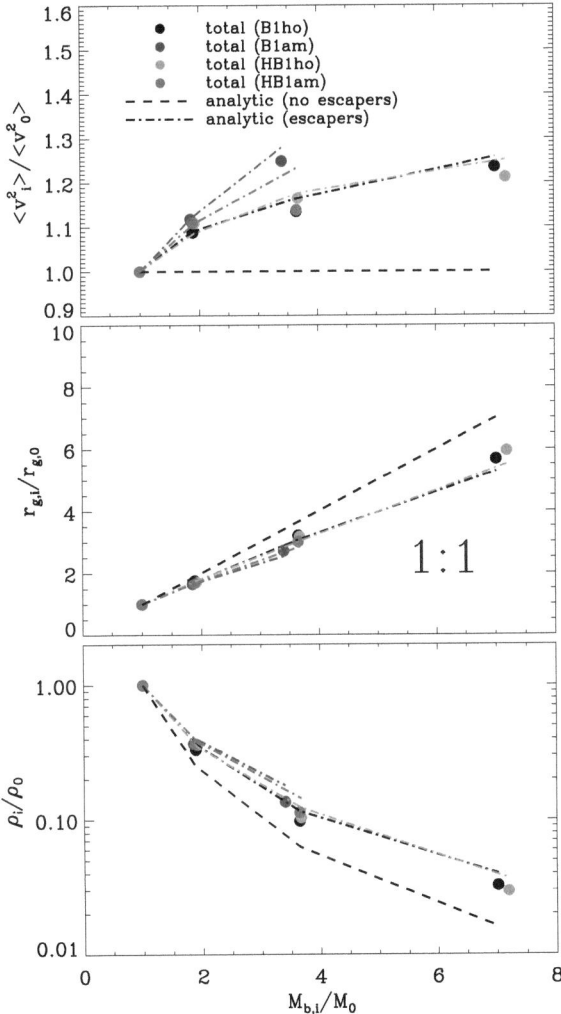

Figure 6.8: Normalized evolution of the mean square speeds (top panel), gravitational radii (middle panel) and densities within the gravitational radius (bottom panel) for all equal-mass merger hierarchies (see table 6.1) plotted against the normalized bound system mass $M_{b,i}$. The dierence in evolution, to a simple analytical estimate (black dashed line, Eqs. 6.7, 6.8, and 6.9) can be accounted for by escapers (dashed-dotted lines, Eqs. 6.15-6.17).

With $\equiv M_{esc}/M_i$ as the ratio of mass, lost in escapers to initial mass and $\equiv \langle v_{esc}^2\rangle/\langle v_i^2\rangle$ as the ratio of the mean square speed of the escapers and the initial system, we can now re-write equations 6.7 to 6.9 as

$$\frac{\langle v_f^2\rangle}{\langle v_i^2\rangle} = \frac{(1+\ \ +\ \ )}{(1+\ \ -\ \ )}, \tag{6.15}$$

$$\frac{r_{g,f}}{r_{g,i}} = \frac{(1+\ \ -\ \ )^2}{(1+\ \ +\ \ )} \tag{6.16}$$

and

$$\frac{f}{i} = \frac{(1+\ \ +\ \ )^3}{(1+\ \ -\ \ )^5}. \tag{6.17}$$

The dashed-dotted lines in Fig. 6.8 indicate that the updated analytic predictions are in good agreement with our simulation results. The deviations are less than a few per cent for the one- and two- component models, respectively.

The situation becomes more complicated, if we separate the velocities of the bulge and the halo component (left panel of Fig. 6.9). The mean square speed of the bulge (green squares) increases more (finally > 50%) with respect to the total system (black squares), whereas the halo (blue squares) speed stays below the total. Here violent relaxation and dynamical friction lead to an energy transfer from the bulge to the halo, i.e. the final bulge is more tightly bound than the initial one (see also Boylan-Kolchin et al. 2005 for a discussion of the eect of dierent orbits).

This eect can be estimated based on the ratio of dark and stellar matter. The total kinetic energy of the system is

$$m_*\langle v_*^2\rangle + m_{dm}\langle v_{dm}^2\rangle = m_{tot}\langle v_{tot}^2\rangle. \tag{6.18}$$

With $m_{tot} \equiv m_* + m_{dm}$ and introducing $\Delta\langle v_{*/dm}^2\rangle = \langle v_{*/dm}^2\rangle - \langle v_{tot}^2\rangle$ we obtain,

$$m_*\Delta\langle v_*^2\rangle + m_{dm}\Delta\langle v_{dm}^2\rangle = 0 \tag{6.19}$$

and the additional growth of the stellar mean square speeds is

$$\Delta\langle v_*^2\rangle = -\frac{m_{dm}}{m_*}\Delta\langle v_{dm}^2\rangle. \tag{6.20}$$

If we now add $\Delta\langle v_*^2\rangle$ to the mean square speed of the galaxy $\langle v_{tot}^2\rangle$ we can consistently predict the bulge dispersion (green solid line in the left panel of Fig. 6.9).

As violent relaxation scatters particles into states with higher binding energy, the central region becomes slightly contracted relative to the totals system growth (depicted by the gravitational radius). In the right panel of Fig. 6.9 we show the radii enclosing the 20, 50 and 80 % most bound particles normalized to the evolution of the gravitational radius. The inner regions expand less and the outer regions more than the gravitational radius. This eect is already described in White (1978, 1979) and

6.4 MAJOR MERGERS

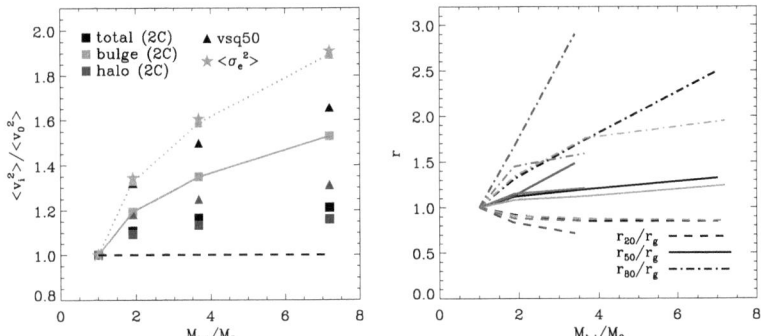

Figure 6.9: Left panel: Here we compare the mean square speeds of the total system (black squares), the halo (blue squares) and the bulge (green squares) for the simulations of two-component models with head-on orbits (HB1ho). The more rapid increase of the bulge velocities can be explained by an energy transfer from the halo to the bulge (green solid line, see Eq. 6.20). In the central regions this eect is even more e cient, as the mean square speeds within the spherical half-mass radius (corresponding triangles) of the bulge increase more. The dashed green line indicates the expectation of Eq. 6.20 for the central region and the green star depicts the mean eective velocity dispersion. Right panel: Evolution of the radii enclosing the 20% (dashed), 50% (solid) and 80% (dashed-dotted) most bound particles normalized to the evolution of the gravitational radius, r_g, (same colors as in Fig. 6.8). The inner regions expand less and the outer regions more than the gravitational radius. All ratios are normalized to the initial values.

also lead to a significant non-homology of the system (see also Boylan-Kolchin et al. 2005), as it evolves through successive mergers.

On the other hand, this high central density leads to higher central velocities (triangles, left panel Fig. 6.9). If we now add $\Delta \langle v_*^2 \rangle$ (Eq. 6.20) to $\langle v_{tot}^2 \rangle$ for the central regions we again get a very good prediction for the central bulge velocity (green dotted line).

Now we change focus from theoretical galaxy properties, like the gravitational radius, to directly observable galaxy properties like the line-of-sight velocity dispersion σ_e, the effective radius r_e and the projected surface density Σ_e. We define r_e as the mean radius including half of the projected bound stellar mass along the three major axis and σ_e, Σ_e as the mean projected velocity dispersions and mean surface densities within r_e, respectively. The green stars in the top panel of Fig. 6.10 show, that the effective velocity dispersion σ_e^2 for the two-component mergers with head-on orbits (HB1ho) evolves similar to the central mean square speeds of the bulge (left panel Fig. 6.9), with a final value a factor ~ 1.9 higher than initial. For the one-component merger hierarchies, the central regions are also affected by the contraction effect and the innermost mass radii grow less than the gravitational radius (right panel Fig. 6.9). This effect is stronger for the scenario with angular momentum orbits, as the final coalescence takes longer and therefore, σ_e^2 increases more compared to the head-on case (top panel Fig. 6.10). That means that the amount of escaping particles, which additionally leads to a contraction of the central regions (right panel Fig. 6.9), changes Σ_e of both systems, but for the two-component mergers there is an additional transfer of kinetic energy from the halo to the bulge particles.

In contrast to the gravitational radii, the observable effective radii of one- and two-component major mergers follow the simple analytic predictions Eq. 6.7-6.9, although the dispersion does not. As $r_e \propto M/\sigma_e^2$ we would expect smaller radii, similar to the evolution of the gravitational radii in Fig. 6.8. Projection effects can be ruled out, as the spherical half-mass radii of the bound remnants evolve similar to r_e and it cannot be an effect of dark matter alone. As discussed in Nipoti et al. (2003) and Boylan-Kolchin et al. (2005) and looking at the different evolution of different mass radii (right panel Fig. 6.9), we also find, that the systems change their internal structure with each merger generation. This effect of non-homology can be quantified by the structure parameter c which connects the stellar mass of the systems to its observed size and velocity dispersion (see also Figs. 6.7),

$$M_* = c \cdot \frac{r_e \sigma_e^2}{G}, \qquad (6.21)$$

where we define M_* as the bound bulge mass of the merger remnants, to get a comparable value for c of both merger hierarchies (see also Prugniel & Simien 1997; Nipoti et al. 2009b). A change in c indicates that the merger remnants do not have a self-similar structure. For the major mergers we find a continuous decrease of c with each merger generation (filled circles in Fig. 6.11). The decrease after one merger generation is slightly stronger (factor of 1.8) for the models including dark matter (red and green

6.4 Major Mergers

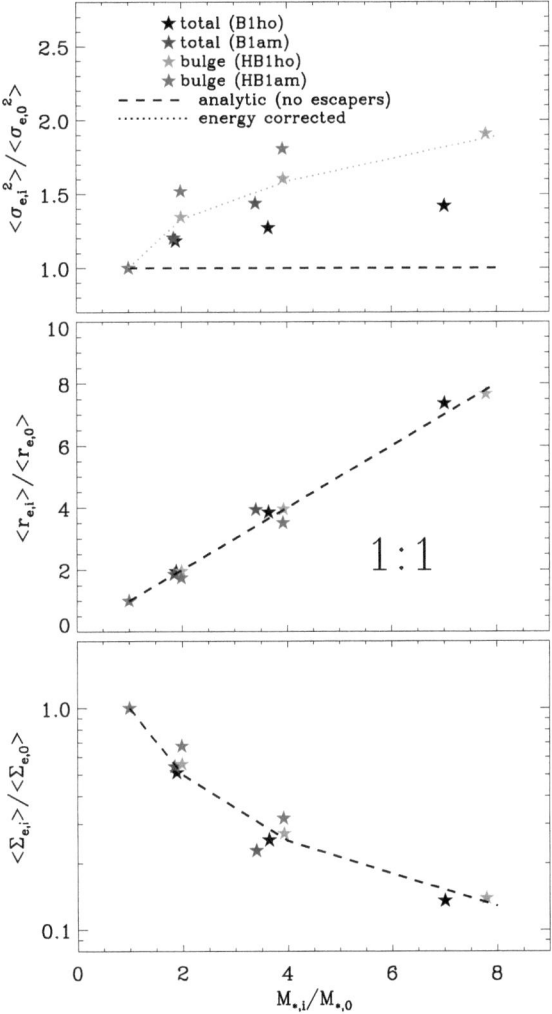

Figure 6.10: Same as Fig. 6.8 but for 'observable' bulge properties like the eective line-of-sight velocity dispersion (top panel), the projected spherical half-mass (eective) radius (middle panel) and the eective sur face density (bottom panel) versus the bound stellar mass normalized to the initial stellar mass. The dashed lines indicate the simple analytic predictions (Eqs. 6.7, 6.8, and 6.9) and the dotted line in the top panel is the expectation for $\frac{2}{e}$ (see also Fig. 6.9). Surprisingly, the observable values for the bulge sizes (and therefore eective surface brightness) agree with the analytic prediction.

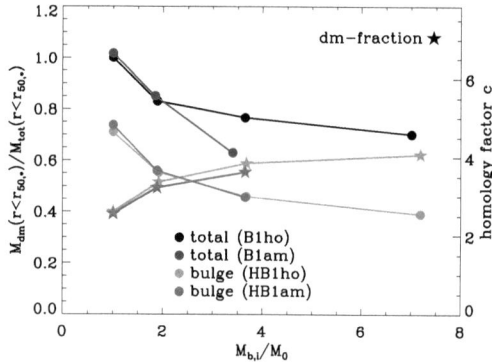

Figure 6.11: The dark matter fraction (stars) within the spherical half-mass radius of the bulge, $f_{\rm dm} = M_{\rm dm}/(M_{\rm dm} + M_*)$ of the two-component major mergers versus the bound mass, normalized to the initial mass. The central dark matter fraction ($r < r_{50,b}$) increases by a factor of 1.5 after three generations for the head-on orbits (green). The filled circles indicate the evolution of the structure parameter $c = GM_{\rm bulge}/(r_e \sigma_e^2)$. The parameter decreases for one-component and two-component systems indicating a break in homology.

circles) compared to the bulge only model with radial orbits (factor of 1.5; black circles). This discrepancy can be explained by the increasing dark matter fraction (stars, Fig. 6.11) within the spherical half-mass radius of the bulge+halo models. In section 6.4.1 we have already shown, that this is not just an eect of increasing radii (as argued by Nipoti et al. 2009b). In the center, the amount of dark matter particles grows more compared to bulge particles and we observe a real increase of central the dark matter fraction. This is in good agreement to Boylan-Kolchin et al. (2005), who also finds that, especially for equal-mass mergers on radial orbits, homology is not preserved and the dark matter fraction increases.

Finally, looking at the scaling relations for the first remnant, we obtain reasonable results compared to, e.g. Boylan-Kolchin et al. (2005), who looked at merger remnants after one generation of equal-mass mergers. The first remnants of our equal-mass merger scenario, gives a mass-size relation,

$$r_e \propto M_*^\alpha, \qquad (6.22)$$

with $\alpha = 0.8 - 1.0$ and a mass-velocity-dispersion relation,

$$M_* \propto \sigma_e^\beta, \qquad (6.23)$$

with $\beta = 3.3 - 5.1$. Compared to observations (e.g. $\beta = 0.56$, Shen et al. 2003), our size increase is too high. But as Boylan-Kolchin et al. (2005) already pointed out, radial

6.5 Minor Mergers

orbits yield an higher growth in size and velocity. As our orbits all have zero or very small pericentric distances, this explains the discrepancies with the observations.

6.5 Minor Mergers

In this section we investigate the eect of minor mergers with initial mass-ratios of 1:10.
In contrast to major mergers, the theoretically predicted size increase per added mass would be higher accompanied with a significant decrease of the velocity dispersions and densities (Eq. 6.7-6.9).

In Figs. 6.12 and 6.13 we show the total energy distributions (solid lines) for a sequence of head-on, one- (B10hoc) and two-component (HB10hod) minor mergers, respectively. For both, nearly all escaping particles are from the satellites (red dashed-dotted line in both figures), which indicates that violent relaxation only aects the in-falling material and has a negligible e ect on the distribution of the host particles. As the satellites are less bound than the host galaxy we get a very high fraction of unbound mass for the final remnants. Furthermore, we can see that almost no accreted particles assemble at the central regions. The shift of the highest bound particles to higher energies (right) in both scenarios is caused by two-body relaxation, which is most eective in these high density regions (see Section 5.1). Combining this shift with the eect, that most satellite particl es assemble at low binding energies, we get an increase of the mean binding energies. For the bulge only mergers, this decrease of the mean energy can also be predicted analytically.

The potential energy for a Hernquist sphere is (Hernquist, 1990),

$$W = -\frac{GM^2}{6a}, \tag{6.24}$$

and according to the virial theorem the total energy of a system in equilibrium is

$$E = \frac{1}{2}W = -\frac{GM^2}{12a}. \tag{6.25}$$

Additionally to the formerly defined mass ratio $\eta \equiv \frac{M_a}{M_i}$ we define the ratio of the accreted and initial scale radii as $\epsilon \equiv \frac{a_a}{a_i}$. Assuming energy conservation for a zero energy orbit, the system's final energy is:

$$\begin{aligned}
E_f &= E_i + E_a = -\frac{M_i^2}{12a_i} - \frac{M_a^2}{12a_a} & (6.26) \\
&= -\frac{M_i^2}{12a_i} - \frac{(\eta M_i)^2}{12\epsilon a_i} = -\frac{M_i^2}{12a_i}\left(1 + \frac{\eta^2}{\epsilon}\right) & (6.27) \\
&= E_i\left(1 + \frac{\eta^2}{\epsilon}\right) & (6.28)
\end{aligned}$$

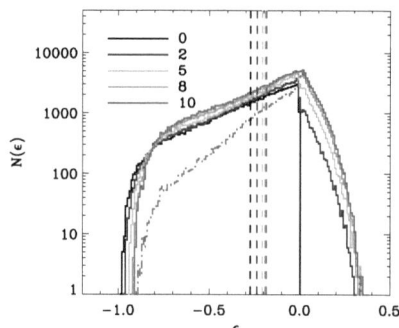

Figure 6.12: Dierential energy distribution for th e initial one-component system (black) and two (purple), five (blue), eight (green), and 10 generations of 1:10 head-on mergers (B10hoc). The red dashed-dotted line indicates the energy distribution of all accreted material, which shows that nearly all escapers come from the satellites and nearly no particles assemble at the center. In contrast to equal-mass mergers (left panel, Fig. 6.4) all bound particles become less bound as the mean binding energy of the systems (vertical dashed lines) decreases with each merger generation.

Furthermore we calculate the mean final energy $\bar{\epsilon}_f$,

$$\bar{\epsilon}_f = \frac{E_f}{M} = \frac{E_i(1+\frac{\mu}{\eta}^2)}{M_h + M_s} = \frac{-\frac{M_i^2}{12a_i}(1+\frac{\mu}{\eta}^2)}{M_h(1+\mu)} \quad (6.29)$$

$$= -\frac{M_i}{12a_i}\left(\frac{1+\frac{\mu}{\eta}^2}{1+\mu}\right) = -\bar{\epsilon}_i\left(\frac{1+\frac{\mu}{\eta}^2}{1+\mu}\right) \quad (6.30)$$

Here we used the fact that for equal mass particles the total number of particles is equivalent to the total mass M. In the case of an equal mass merger of two identical systems $\mu = \eta = 1$ and $\bar{\epsilon}_f = \bar{\epsilon}_i$, i.e. the mean energy of the system stays constant (see also top panels Figs. 7.3, 7.5). But for the numerical setup of the first minor merger generation B10hoc (B10hod), where $\mu = 0.1$ and $\eta = 0.5 (1.0)$, the final mean energy is $\bar{\epsilon}_i / \bar{\epsilon}_f = 0.92 (0.93)$, in agreement with the simulations (Fig.6.12).

Taking a closer look on the energy distribution of the bulge of the scenario HB10hod (dashed lines, top panel Fig. 6.13) we can directly see, that most stellar particles accrete at energies $\epsilon > -0.4$, creating an overdensity of bulge particles. Consequently the ratio of dark matter particles to bulge particles decreases for $\epsilon > -0.4$ (bottom panel of Fig. 6.13). In the latter panel we can also see, that this ratio stays constant for all particles with $\epsilon < -0.4$ and as the binding energy of the 50% most bound particles go to higher energies, the dark matter matter fraction increases (see also Fig.6.16).

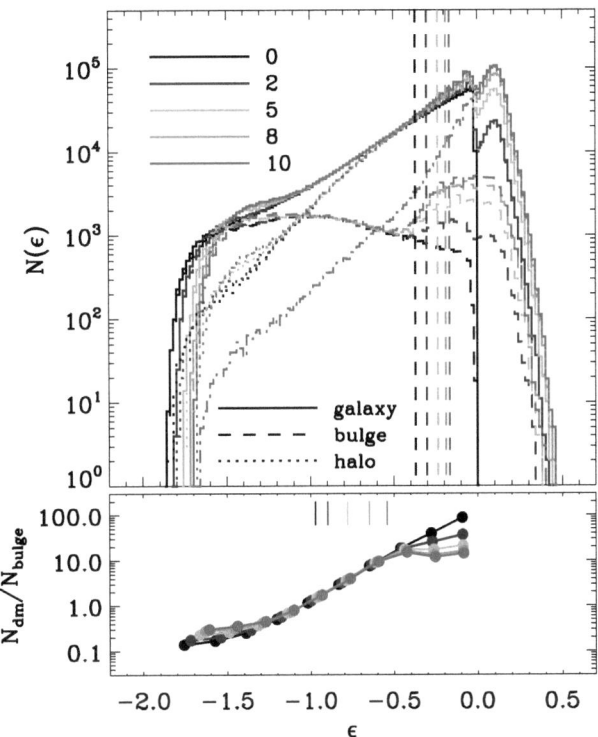

Figure 6.13: Top panel: Dierential energy distribution for the initial two-component system (black) and two (purple), five (blue), eight (green), and 10 generations of 1:10 mergers (HB10hod). The red dashed-dotted line shows that all particles with positive binding energies are from the satellites, which means that violent relaxation only unbinds satellite particles and the energy distribution of the host stays unaected. In contrast to equal-mass mergers (Fig. 6.5) bound particles become less bound, due to two-body relaxation. As for the one-component model, the mean binding energy of the system (vertical dashed lines) decreases with each merger generation. Bottom panel: The relative fraction of dark matter particles and stellar particles at low binding energies remains unchanged and decreases for > -0.4. The short vertical lines at the top of this panel indicate that the energy of the 50% most bound bulge particles gets shifted to higher lower binding energies.

6.5.1 Velocity dispersion

Next we focus on the evolution of the velocity dispersion profiles (Fig. 6.14). Regarding the dierential energy distribution (Figs. 6.12, 6.13) we have seen, that violent relaxation has no big influence on the central regions of the host galaxies. In Fig. 6.14 we can see, that this is also reflected in the radial velocity dispersion profiles $\sigma_r(r)$ of the one- and two-component minor mergers. Especially in the case of the one-component scenario (e.g. B10hod in the top left panel) $\sigma_r(r)$ keeps the initial Hernquist profile (black dashed line) over the whole radial range. Checking the other one-component scenarios, we also found, that even after 10 accretion events, the velocity profile does not change. Furthermore, as $\beta(r) = 0$ (bottom left panel Fig. 6.14) all remnants (solid lines) stay perfectly isotropic. This picture changes, if we just look at the accreted material, which approaches on radial orbits and thus shows growing radial anisotropy $\beta(r) > 0$ with increasing radius. Again this eect is the same for all one-component minor mergers, but it is slightly less pronounced in the case, where the satellite's orbit has some angular momentum.

The merger remnants of two-component models also indicate characteristics, which are consistent with the evolution of their dierential energy distribution (Fig. 6.13). Therefore, the total velocity dispersion profile (solid lines in the right panel of Fig. 6.14) and the one of the halo stay constant. But as we have already seen, a lot of stellar particles create a bump in the energy distribution (dashed lines in Fig. 6.13), which gets more and more prominent with each subsequent generation. These accreted particles induce an increasing velocity dispersion at radii larger than the spherical half-mass radius r_{50} (dotted lines, right panel Fig. 6.14). As the final coalescence of the stellar component in the 2C scenario is on radial orbits, independent of the initial conditions (see also González-García & van Albada (2005)), the anisotropy parameter gets radially biased for all of our minor mergers (e.g. HB10hod, bottom right panel of Fig. 6.14). This eect only occurs in the simulation including a dark matter halo, because then the angular momentum of the in-falling satellite gets lost before the final merger due to enhanced dynamical friction. Hence, most of the stellar particles approach on radial orbits and get, during the final coalescence, stripped before reaching the center. If we use the compact satellites, the overall trend does not change, but more material gets closer to the center, as the particles are more tightly bound and suer less from tidal stripping.

6.5.2 System Evolution

The evolution of the total bound minor merger generations are depicted in the left panels of Fig. 6.15. Obviously the mean square speeds (top panel) of all hierarchies decrease with increasing mass. In all scenarios with diuse satellites (black, blue, green and red filled circles), the evolution is very close to the virial expectations of Eqs. 6.7-6.9 (dashed line), although the mass loss is significant especially for the two-component models (red and green circles). In table 6.1 we can see that the fraction

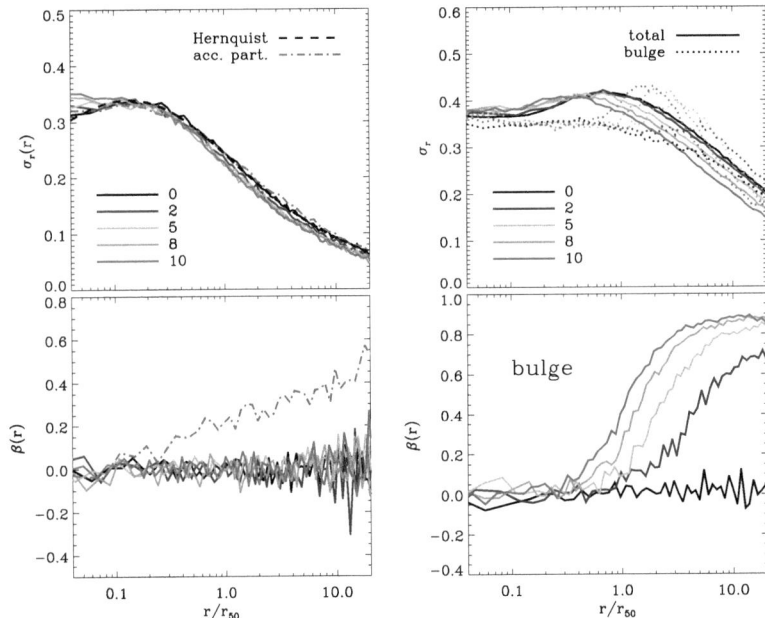

Figure 6.14: Top panel: The radial velocity dispersion for the head-on minor mergers of one-component models (B10hoc) stays constant over most of the radial range. Only in the very central regions, it increases slightly with each generation. The black dashed line is the initial Hernquist profile and the red dashed-dotted line the velocity dispersion of all bound accreted particles. Bottom panel: For the whole bound remnant, the velocity distribution stays perfectly isotropic, as the anisotropy parameter stays zero. Looking at the accreted material (red dashed-dotted line), it gets radially anisotropic with increasing radius. In both panels the radius is normalized to the spherical half-mass radius of the bound system.

Top panel: The radial velocity dispersion of the total system (solid lines) for the head-on minor mergers of two-component models (HB10hod) stays constant over the whole radial range. The dispersion of the bulge system (dotted line) builds up a prominent bump which comes from the accreted material, that gets stripped in the outer parts of the host system. The radii are normalized to the spherical half-mass radius of the bulge. Bottom panel: The anisotropy parameter of the bulge velocities gets radially biased at radii greater than the spherical half-mass radii of the bulge.

of escaping particles is up to 35% for HB10hod and more than 20% for the other scenarios. Furthermore, regarding the 2C models, most of the escape fraction is due to the dark matter particles. Going back to the evolution of $\langle v^2 \rangle$, we can see, that the corrected prediction of Eq. 6.15 (dashed-dotted line), which includes the eect of mass loss, perfectly fits the results (e.g. scenario B10hod). Using more compact satellites the final decrease of velocities (orange and purple circles) is much weaker, because they are more tightly bound. As they have half the scale radius of the diuse satellites, their binding energies and velocities are two times higher which then doubles the velocity fraction $= \langle v_a^2 \rangle / \langle v_i^2 \rangle$ of Eqs. 6.7-6.9 and yields a smaller decrease. In combination with the occurring mass loss, this explains the dierent evolution of the mean square speeds. Nevertheless, in all scenarios the final mean square speeds of the total systems are $10-30\%$ lower compared to their initial host galaxies, which is in good agreement to observations, that predict a mild decrease of the compact early-type's velocity dispersions.

The evolution of the gravitational radii (middle left panel of Fig. 6.15) of the six hierarchies evolve according to the mean square speeds, which is not surprising as $r_g \propto 1/\langle v^2 \rangle$ (see Eq. 6.5). In detail, this means, that the hierarchies with a diuse satellite show a size increase, which is consistent with the analytic predictions (dashed line) and as the compact satellites are not able to e ciently decrease the velocities, their gravitational radii grow only marginally. However, for all minor mergers the maximum size growth is around a factor ~ 2.4, which is by far too weak to explain the observed evolution of compact early-type galaxies. For completeness, the bottom left panel illustrates, that the mean density within the gravitational evolves according to the gravitational radius ($\propto r_g^{-3}$).

In the right panels of Fig. 6.15 we illustrate the eective line-of-sight velocity dispersion $_e$ (top), the eective radius r_e (middle) and the eective surface density of all minor merger remnants. Obviously, the central regions show nearly no evolution of $_e^2$, except the two bulge only scenarios with a diuse satellite (B10amd, B10hod). Before we explain the dierent results, we fir st look at the size evolution of the according eective radii r_e (middle panel) and the eective surface densities (bottom panel). Surprisingly, the sizes of nearly all merger remnants grow significantly and for the most e cient one (HB10hoc) the final size is a factor 4 .5 higher, which is even much higher than the virial expectation (Eq. 6.8). This strong evolution is also reflected in the eective surface densities (bottom panel), which decrease at maximum by an order of magnitude.

In the case of bulge only scenarios, the dierent evolutions of the central parameters can be explained by structural changes, measured by the structure parameter c (Eq.6.21). In Fig. 6.16 the filled circles show, that the three one-component minor mergers indicate dierent results. The B10hoc sequence evolves nearly self-similar, i.e. it grows at all radii and finally does not change its initial shape (see also Fig. 6.12). Consequently, the total system evolves the same as the central system and the increase of the eective radius is very similar to the gravitational radius (left middle panel). Furthermore, due to the escapers, the do not grow notably, thus the calculation of the

6.5 Minor Mergers

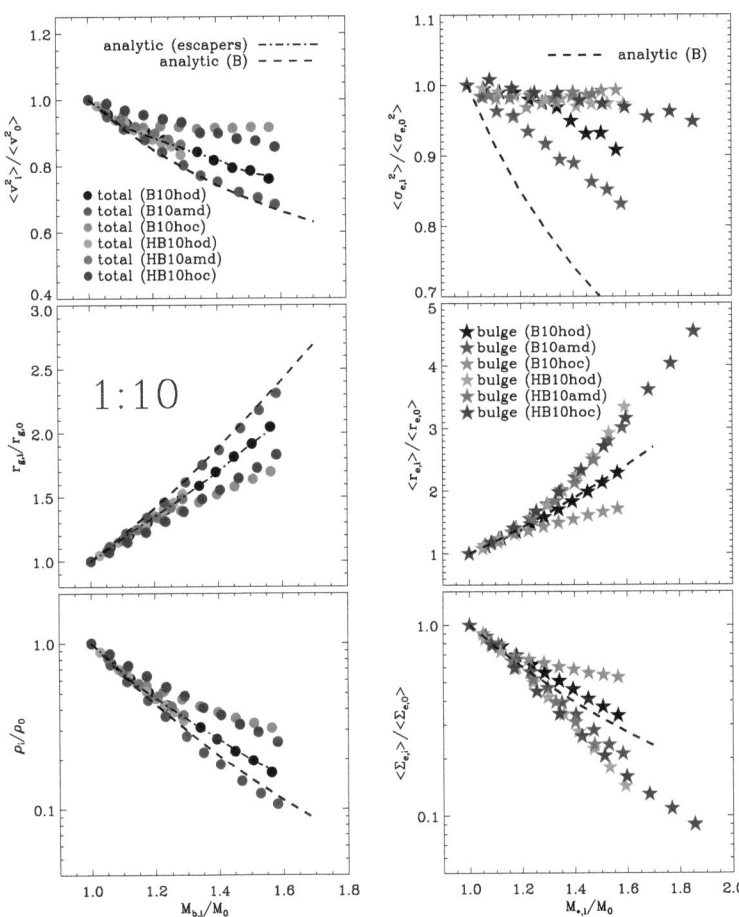

Figure 6.15: Left panels: Evolution of the mean square speeds (top), the gravitational radii (middle) and spherical densities within r_g (bottom) for all minor merger scenarios (see table 6.1). The dashed lines in each panel are the idealized expectations of Eqs. 6.7-6.9 for the all diuse one-component scenarios a nd the black dashed-dotted line depicts the corrected expectations of Eqs. 6.15-6.17 for the minor merger scenario B10hod.

Right panels: The squared mean line of sight velocity dispersion (top), the mean eective radius (middle) and the mean eective density (b ottom) for the scenarios of the left panels. In contrast to the total system, the central velocity dispersion shows nearly no decrease, except for the hierarchy B10amd, but a very high size increase. Only B10hoc, with a compact satellite stays below the idealized expectations (dashed line). Here, the x-axis indicates the stellar masses of each remnant and not the total system masses.

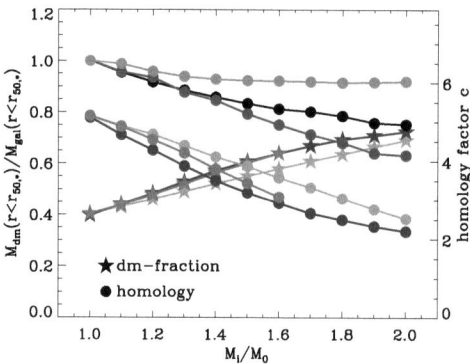

Figure 6.16: The left y-axis and the stars show an increasing dark matter fraction within the spherical half-mass radius of the bulge for the two-component minor mergers. The right y-axis together with the circles indicate a strong decrease of the structure parameter c (eq. 6.21) for nearly all minor merger scenarios. Due to the high mass loss of some scenarios we plot all values against the total system mass. Colors are the same as in Fig. 6.15.

eective line-of-sight velocity dispersion is restricted to the central parts, with high velocities, and therefore stays constant. The further two bulge only scenarios, both include weakly bound satellites, which already loose most of their material in the outer regions of the host galaxy. Hence the latter ones build up an extended envelope, while the centers stay unaected, i.e. the structural properties of the remnants do change (see also black and blue circles in Fig. 6.16). On the other hand, the development of an extended envelope boosts the size growth of a system. As the the sequence B10amd (blue circles) with an angular momentum orbit needs more time until the final coalescence, it suers more from tidal stripping and builds up the most extended envelope of all bulge only models, which then results in the highest size growth. This implies, that the calculation of σ_e also includes particles outside the innermost regions, where the velocities are lower and the velocity dispersion within the eective radius decreases (see also top right panel Fig. 6.14).

Regarding the evolution of the bulge+halo scenarios we additionally have to deal with the eect of dark matter, which also has a big influence on the evolution of the observable properties. In the middle panel of Fig. 6.15 we can see, that all three scenarios yield a significant size growth up to a factor of ~ 4.5 (HB10hoc), which is the consequence of a developing extended envelope. In Fig. 6.17 we illustrate the evolution of the surface density along the major axis. Obviously, most of the accreted material settles down at larger radii $r > 10$ and does not reach the center, which directly highlights the build up of the stellar envelope and the structural change of the

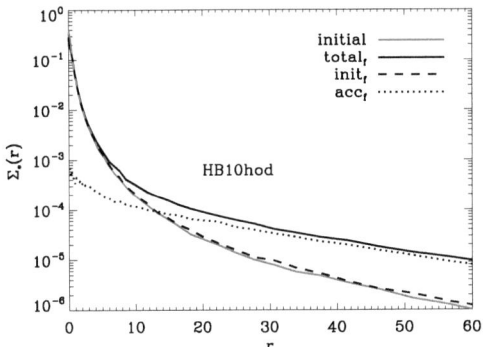

Figure 6.17: Surface densities of the bulge along the major axis for the head-on minor mergers of two-component models (HB10hod). The grey solid line indicates the initial host surface density, which is the same as the final surface density of the host particles (black dashed line). Most of the satellite's material (dotted line) assembles at a radius $r > 10$ and increases the final profile (black solid line) especially in the outer parts, while the central profile stays the same.

final remnant. As the final structure parameter is very similar for all two-component minor mergers (green, red and purple circles in Fig. 6.16), they all follow the same evolutionary path with respect to the size growth. In the case of scenario HB10hoc, the satellite is a factor 2 ore bound, which induces two consequences, first, some particles go slightly further to the host's center and second, less mass is lost during the merger process, which results in the most e cient size growth. So far, dark matter enhances tidal stripping and leads to the build up of an extended envelope, regardless of which orbit we use. But as the radius increases that rapidly, the eective radius goes into regions which are more and more dark matter dominated, which finally results in a highly increasing dark matter fraction (stars in Fig. 6.16). In the end the ratio of initial to final dark matter mass within the spherical half-mass radius is a factor of > 1.8 higher. But, contrary to the equal-mass mergers, this increase is just a result of the size growth as the real fraction of bulge to halo particles do not change over most of the energy space (see bottom panel Fig. 6.13). Additionally, the increasing dark matter fraction within the half-mass radius keeps the velocities of stellar particles constant out to a much larger radius compared to the bulge only models (see also top right panel Fig. 6.14). Therefore the eective line-of-sight velocity dispersions in the top right panel of Fig. 6.15 do not change.

Altogether, we can say that minor mergers are very e cient drivers for the size growth of spheroidal galaxies. As dark matter enhances dynamical friction and tidal stripping, it enhances the eect and due to the finally high eective radii, the dark matter fraction also grows by nearly a factor of 2 after 10 generations of minor mergers.

6.6 Summary & Discussion

We have performed numerical simulations of frequent major and minor mergers of spherical, isotropic galaxy models, which consist of one- and two-component Hernquist spheres. After testing the models for their stability we performed two and three generations of equal-mass mergers on orbits with and without angular momentum of either one- or two-component models. The main results can be summarized as follows:

- During an equal-mass merger, violent relaxation plays an important role. First, it leads to non-negligible amount of mass-loss and second, the dierential energy distribution goes to much higher binding energies.

- Violent relaxation and mixing leads to a 'real' increase of the central dark matter fraction, as more dark matter than stellar particles are mixed into the center.

- Due to phase mixing, violent relaxation vanishes rapidly and therefore never reaches its final state of an isothermal sphere. But with each subsequent equal-mass generation we get closer to the state of maximum entropy and the final velocity dispersion profile can well be fitted by a Jae profile (Jae , 1983), which resembles the inner parts of an isothermal sphere.

- All merger remnants get radially anisotropic velocities, as we only use radial or close to radial orbits.

- This aects the merger remnants in a way, that the mean square speeds of the total systems increase significantly and the size growth is less than expected from theoretical predictions, which ignore escaping particles.

- As the central binding energies increase significantly, the central velocities and the LOSVD increase even more compared to the total system.

- Including dark matter enhances dynamical friction which is able to transfer energy from the bulge to the surrounding halo and increases the central bulge velocities even more than in the one-component scenario.

- Due to a strongly decreasing structure parameter, homology breaks and the eective radii of the remnants evolve exactly like the theoretical expectation ($r_e \quad M \quad$).

In the case of minor mergers, the initial mass ratio is assumed to be 1:10 and we used orbits with and without angular momentum. As initial satellite galaxies we take two extreme cases, i.e. they are either very diuse or very compact. The main results for the minor mergers are:

6.6 Summary & Discussion

- Violent relaxation does not effect the overall differential energy distributions of the host galaxy.

- Due to dynamical friction and tidal stripping the stellar particles develop a prominent bump at low binding energies.

- The velocity dispersion of the bulge only (one-component) models do not change their shape, keep their initial Hernquist profile and stay isotropic over the whole radial range.

- For two-component accretions the final coalescence of the bulges always is on radial orbits, the stellar velocities become radially anisotropic at radii approximately larger than the spherical half-mass radius

- Using diffuse satellites, the mean square re speeds of the remnants decrease with each subsequent generation, which is only limited by the high amount of mass-loss and consequently the gravitational radii increase much less than expected.

- The head-on minor mergers of compact one-component models evolve nearly homologous, i.e. the observable values like the line-of-sight velocity dispersion and the effective radius evolve very close to those of the whole system.

- In all other minor merger sequences, we observe a dramatic break of homology, as the remnants build up an extended envelope of stars, while the central configuration stays constant.

- Therefore the effective radii increase rapidly up to a factor of 4 .5, which is much closer to virial expectations.

- The rapid size growth results in a significant increase of the dark matter fraction within the spherical half-mass radius up to a factor of ~ 1.8.

- Due to the increasing dark matter fraction, the effective line-of-sight velocity dispersions do not decrease but stay constant.

One important question which has to be solved for elliptical galaxies is, how the compact early-types at a redshift $z \sim 2$ grow with time. As their stellar distribution is already red without significant star formation, we used dry mergers to explain this evolution. van Dokkum et al. (2010) finds a size-mass relation of $r_e \propto M^{2.04}$, which indicates a size increase of a factor of 4 as the galaxy's mass gets doubled since $z \sim 2$. The resulting relation of our minor merger scenarios of two-component models is even higher $r_e \propto M^{>2.04}$ up to a exponent of 2.4, which shows that dissipationless minor mergers are a good way to solve this problem. However, Nipoti et al. (2009a) tried a similar approach and find a much lower size increase in their simulations ($r_e \propto M^{1.09}$). One reason for this big discrepancy is, that they calculated the exponent of the stellar mass by averaging over all their merger hierarchies. As they have more major mergers than minor mergers, this of course lowers the size increase significantly. Additionally, they use a steeper slope for the stellar density profile of the host and the satellite galaxies, where the size increase can not be that e cient, as the accreted material is more concentrated in the satellite's center compared to our setup. Finally, their satellites are even more compact than our satellites which lie on an extrapolation of the $z = 2$ mass-size relation of Williams et al. (2010).

Furthermore we find that the dark matter fractions for our idealized simulations agree well with previous work, where the dark matter fraction increases in dry mergers. This changes the ratio of dynamical and stellar mass and might, e.g. help to explain the tilt of the fundamental plane (Boylan-Kolchin et al., 2005). Of course, that is just one possibility to explain the tilt and Grillo & Gobat (2010) suggest that it depends more on M_*/L, but it is not clear yet how strong the single contributions are. We also agree with Nipoti et al. (2009a), that the increase of the dark matter fraction is more e cient for minor mergers and for this scenario is dominated by the rapid size growth. But in contrast to Nipoti et al. (2009a), we find that the central dark matter fraction of equal-mass mergers illustrates a 'real' change caused by violent relaxation and mixing.

Looking at the velocities at dierent radii, our minor merger results are not able to explain recent observations of very high velocity dispersions at high redshift (van Dokkum et al., 2009; van de Sande et al., 2011). Our results indicate, that we get a decrease of the mean square speeds of the total system, but the observed line-of-sight velocity dispersion hardly changes. This indicates, that simple dissipationless mergers are not able to decrease the very high LOSVD of some compact early type galaxies (van Dokkum et al., 2009). This problem might be solved, if we include some gas and AGN feedback or use more realistic galaxy models, which have dierent orbital properties.

But altogether our work shows that dissipationless dry mergers are able to increase the size of a compact early type galaxy. As we lie even above the observed predictions a small amount of gas, which is known to lower the size growth (Covington et al., 2011; Hopkins et al., 2008), would perhaps not be enough to destroy this scenario.

CHAPTER 7

SIZE AND PROFILE SHAPE EVOLUTION OF MASSIVE QUIESCENT GALAXIES

In this chapter, we focus on the evolution of the density structure and the size evolution of compact early-type galaxies and try to understand the importance of dark matter. We know that the sizes and mass distributions of compact, quiescent, massive galaxies evolve rapidly from $z \sim 2-3$ to the present. Many of the $\sim 10^{11}$ systems at high redshift have sizes of $\sim$1kpc and surface brightness profiles with Sersic indices < 4. At $z = 0$ elliptical galaxies above $2 \cdot 10^{11}$ solar masses are more than a factor of 4 larger, indicating a size evolution of r $\propto M$ with ≥ 2. They also have surface brigtness profiles with $n_{\rm ser} \geq 8$. Within a hierarchical galaxy formation scenario this evolution can be explained under two assumptions. The galaxies predominantly grow by mergers with lower mass galaxies and the galaxies have to be embedded in massive dark matter halos so that stars of merging satellites are stripped at large radii increasing the profile shape parameter. We draw these conclusions from idealized simulations of the growth of compact spheroidal galaxies - with and without dark matter - by repeated collisionless mergers with mass ratios of 1:1, 1:5, and 1:10. In simulations without dark matter the sizes evolve less than the corresponding bulge+halo scenarios. If the galaxies are embedded in dark matter halos the stars of the lower mass satellites are more e ciently stripped at large radii resulting in a significantly faster size increase than expected from virial estimates. Repeated 1:5 mergers give $= 2.3$ and after only two merger generations the Sersic index has already increased to $n_{\rm ser} > 8$. For an assumed mass increase of the observed galaxies of a factor of two since z =2 we conclude that the presence of a massive dark matter halo around the galaxies during their minor merger driven assembly is necessary to explain simultaneously their large present day sizes, r > 4 kpc and high Sersic indices, $n_{\rm ser} > 6$.

7.1 Introduction

In the currently favored cosmological CDM model, the universe consists of 24% matter and 76% dark energy ( ), where only 4% of the total matter is in baryonic form (e.g. Spergel et al. (2007)). The other 96% consist of cold dark matter, which has not been detected directly, but has been most successfully applied to explain many observational caveats like the rotation curves of spiral galaxies. On large scales the CDM model, shows very good agreement with observations of the cosmic microwave background and the large scale structure of galaxies. In the context of the CDM model, structure in the universe forms bottom up (White & Rees, 1978; Davis et al., 1985), where the first objects collapse at high redshifts due to fluctuations in the background density field. These first objects merge and build up the dark matter halos of today's observed galaxies.

The baryons assemble in the potential wells of these dark matter halos and form stars which build the observable parts of the universe. The brightest and most massive objects are elliptical galaxies, which form at a redshift of $z \propto 2-3$ in gas-rich major disk mergers (Davis et al., 1985; Bournaud et al., 2011) and due to giant cold gas flows, directly feeding the central galaxy (Kereš et al., 2005; Naab et al., 2007, 2009; Joung et al., 2009; Dekel et al., 2009; Kereš et al., 2009; Oser et al., 2010). Their subsequent evolution is not fully understood yet, as these ellipticals are a factor $\sim$ 4-5 smaller than their counterparts in the present day universe. On the other hand, they are already quiescent, without star formation, and are only a factor of $\sim$2 less massive (Daddi et al., 2005; Trujillo et al., 2006; Longhetti et al., 2007; Toft et al., 2007; Zirm et al., 2007; Trujillo et al., 2007; Zirm et al., 2007; Buitrago et al., 2008; van Dokkum et al., 2008; Cimatti et al., 2008; Franx et al., 2008; Saracco et al., 2009; Damjanov et al., 2009; Bezanson et al., 2009). In addition, they have very dierent surface brightness profiles. In particular, the compact, high redshift ellipticals always have steep power law cusps in their center whereas the more extended present day ellipticals have cored profiles. This means, that fitting a Sersic profile to elliptical galaxies either results in central extra light, where the central surface brightness is above the fitted profile or in case of core ellipticals the profile predicts more light than the galaxy has. The Sersic indices, which are a measurement of the profile's curvature are $\sim$4 for the cuspy galaxies and $\sim$8-10 for the core ellipticals. Furthermore, recent observations of strong gravitational lensing in the SLACS sample (Koopmans et al., 2006; Bolton et al., 2008; Gavazzi et al., 2007, 2008; Auger et al., 2009, 2010) have revealed an increasing central dark matter fraction with stellar mass (Barnabè et al., 2011).

In this chapter, we investigate the evolution of elliptical galaxies with the aid of high-resolution N-body simulations of idealized one- and two-component galaxy models. With dierent initial mass ratios and a dierent choice of merger orbits we explore the eect of frequent dissipationless galaxy mergers. In section 7.2 we give a short overview of the galaxy properties and the simulation parameters. In Section 7.3 we present the eciency of dry mergers for the size growth of compact galaxies and in Section 7.4 we look at the evolution of the surface densities and the mass assembly in multiple merger generations. In Section 7.5 we convert our surface densities to viable surface brightness profiles and explore the evolution of the Sersic profile and in Sections 7.6 we illustrate the change of dark matter fractions. Finally we draw our conclusions in Section 7.7.

7.2 Simulations

We extend a set of simulations of dissipationless mergers of spheroidal galaxies with and without dark matter halos and with mass ratios of 1:10 and 1:1, presented in Chapter 6, by a new set of simulations with an initial mass-ratio of 1:5. We refer to Chapters 4 and 6 for the details of the generation of the stable initial conditions and briefly summarize the simulations setup here. As initial galaxy models, we use isotropic, spherical symmetric one- and two-component models which have a Hernquist density profile (Hernquist, 1990) either for a bulge-only model (one-component, 1C) or a bulge embedded in a Hernquist dark matter halo (two-component, 2C). For the latter case we assume a dark matter to stellar mass ratio of $M_{dm}/M_* = 10$ and the ratio of the scale radii is $a_{dm}/a_* = 11$ The host galaxies have a stellar mass of $M_{*,host} = 1$ and a scale radius of $a_{*,host} = 1.0$. Our satellite galaxies are very diffuse and have for both minor merger scenarios the same scale radius $a_{*,sat} = 1.0$ and are initially 5 or 10 times less massive. For a better comparison to observations we chose a mass scale of $M = 10^{11} M_\odot$ and a length scale of $r = 0.55 kpc$, which yields $v = 884 kms^{-1}$ and $t = 6.12 \times 10^5 yr$. In Fig. 7.1 we can see the positions of our initial galaxies, compared to the most recent, observed mass-size relations. Of course, as we want to investigate the evolution of a compact early-type elliptical, our host galaxies (black filled circle) are below the relation of Williams et al. (2010) at a redshift of $z \approx 2$ (red solid line) and as the satellites are very diffuse (red and green filled circles), they are above an extrapolation of this line. Therefore we also made comparison runs, where the satellites would fall on the high redshift estimates (open circles), i.e. for the mass ratios of 1:5 and 1:10 the satellites have $a_{*,sat} = 0.8 \times a_{*,host}$ and $a_{*,sat} = 0.5 \times a_{*,host}$, respectively.

In the case of equal-mass mergers we simulated two merger generations of both galaxy models. The first generation was a parabolic merger of the galaxies represented by the initial conditions. The second generation was a parabolic re-merger of the duplicated, randomly oriented, first generation merger remnant, which was allowed to dynamically relax at the center. The simulations were performed on orbits with angular momentum and a pericentric distance of one-half the bulge's spherical half-mass radius of the progenitor remnants. Therefore the pericentric distances increase with each merger generation.

The sequences of minor mergers with initial mass-ratios of 1:5 (1:10) were also simulated with one- and two-component models. Initially, the mass-ratio was 1:5 (1:10) and the galaxies were set on parabolic orbits. The randomly oriented merger remnants of the first generations were then set on a parabolic orbit with the initial satellite galaxy models and a mass-ratio of now 1:6 (1:11), and so on. We performed 6 generations of 1:10 mergers and 5 generations of 1:5 mergers using the diffuse satellites (Sat 1:5 (1:10),

7.2 SIMULATIONS

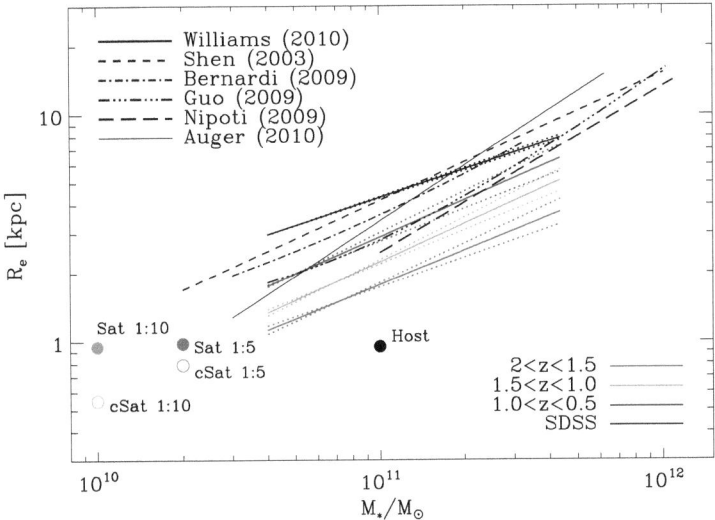

Figure 7.1: The black lines show different observed mass-size relations of the present-day universe. The thick solid lines are the estimates of Williams et al. (2010) for different redshift bins. The corresponding dotted lines are the errors of the latter relations. The filled circles give the position of our compact host (black) and the diffuse initial satellites of the 1:5 (red) and 1:10 (green) minor merger hierarchies. The open circles show the more compact satellites, which would lie on an extrapolation of the $z = 2$ mass-size relation (red line).

see Fig. 7.1). For both scenarios, we set the pericentric distances to half the spherical half-mass bulge radius of the massive progenitor galaxy. In the comparison runs, where the satellites are more compact(cSat 1:5 (1:10), see Fig. 7.1) we computed 4 generations for the bulge+halo models with the same orbits as before, i.e. the pericentric distances are again the half-mass radii of the massive progenitor. In the bulge only scenario with compact satellites we performed a full set of hierarchies, e.g. 5(10) generations for the mass ratios of 1:5 (1:10). In the 1:10 scenario, we also add a hierarchy with 10 generations of head-on minor mergers using a compact two-component satellite (see also Chapter 6).

Looking at the time-scales, we find for our choice of physical scaling, that the longest set of simulations (the 1:10 bulge only) takes $\approx$ 9Gyr, which is very close to the lookback time of $\sim$ 10Gyr ($z = 2$), but all other scenarios are completed in less than $\approx$ 7Gyr. As the 1:10 bulge only scenario has no dark matter halo, the in-falling satellites suer less from dynamical fricti on, and the final coalescence takes by far the longest time.

7.3 Size Evolution

After the completion of every merger, we allow the central region of the remnant to relax, before we compute the projected circular half-mass radii, r_e, along the three principal axes and the bound stellar mass, M_*. In Fig. 7.2 we show the evolution of the mean value of the half-mass radius along the three principal axes as a function of the bound stellar mass for 1:1 (blue), 1:5 (red), and 1:10 (green) merger hierarchies. The black line shows the observed evolution, $r_e \propto M^{2.04}$, in the mass-size plane from $z \approx 2$ to the present day (van Dokkum et al., 2010). The shaded area beyond this line indicates the region, where the size growth per added mass is too small to explain the evolution of compact early type galaxies.

Equal-mass mergers show an almost linear increase of size with mass, (see also Hilz et al. 2011 in prep., Boylan-Kolchin et al. 2005; Bezanson et al. 2009; Nipoti et al. 2009b), independent of whether the stellar system is embedded in a dark matter halo or not (blue solid and dashed lines). As discussed in Boylan-Kolchin et al. (2005), in mergers with dark matter halos the in-falling galaxy suers more from dynamical friction in the massive dark matter halo of the companion galaxy, resulting in more energy transfer from the bulge to the halo. This leads to a more tightly bound bulge with a smaller size (blue solid line, Fig. 7.2) compared to the model without dark matter (blue dashed line, Fig. 7.2). If we combine the results of both major mergers this yields a mass-size relation of $r_e \propto M^{0.91}$ which is similar to the results of Boylan-Kolchin et al. (2005), who found a smaller exponent (≈ 0.7) for orbits with high angular momentum and an exponent of > 1 for pure radial orbits. Nevertheless, as the size grows only linearly with mass, dissipationless major mergers cannot be the main drivers for the subsequent size evolution of compact early-type galaxies.

As expected, from simple virial estimates (Naab et al., 2009; Bezanson et al., 2009),

7.3 Size Evolution

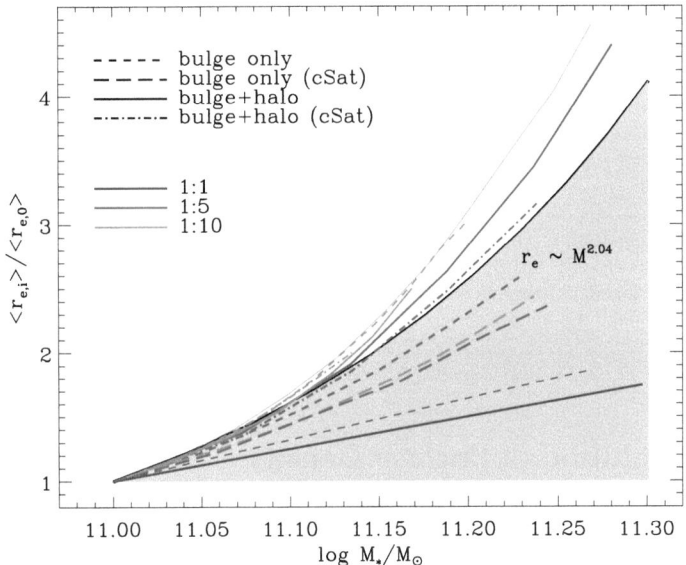

Figure 7.2: The projected spherical half-mass radius (the mean value along the three principal axes) as a function of bound stellar mass for 1:1 (blue), 1:5 (red), and 1:10 (green) mergers. The observed size growth is indicated by the solid black line (van Dokkum et al., 2010). The size evolution of models in the grey shaded area is too weak to be consistent with observations. All mergers of bulges embedded in massive dark matter halos and high mass-ratios (1:5, 1:10, red and green solid/dashed-dotted lines) show a rapid enough size evolution. The size evolution of the bulge-only models (short and long dashed lines) are not e cient enough, except the 1:10 scenario with a diuse satellite (green dashed line). Additionally, we can see, that the accretion of compact satellites leads to less size growth compared to the diuse satellites, but for all bulge+halo scenarios it is still high enough with $r_e \propto M^{>2.1}$. After ten generations of 1:10 head-on mergers with compact satellites (thin solid line), the size increases by a factor of ≈ 4.5, which is more than enough compared to observations.

the size evolution is stronger for bulge-only models with higher mass-ratios of 1:5 and 1:10. This is in good agreement with our simulations, as all red (1:5) and green (1:10) dashed lines have a much larger size increase per added mass, compared to the equal-mass mergers (blue lines). However, except the 1:10 mass ratio with a diuse satellite (green short dashed line), all minor mergers with bulge-only satellites are not e cient enough to lie above the 'forbidden', grey shaded area, which indicates a too weak size growth, compared to observations (van Dokkum et al., 2010).

This picture improves for minor mergers of two-component models, where bulges are embedded in a massive dark matter halos. For all mass-ratios (1:5 and 1:10) and models, i.e. for diuse (solid lines) as wel l for compact (dashed-dotted) satellites, the size evolution is in excess to the observed evolution. In the case of 1:5 minor mergers, with a less compact satellite (red solid line) and for the head-on hierarchy with a compact satellite (thin green line), we get a mass-size relation of $r_e \propto M^{2.28}$ and $r_e \propto M^{2.45}$, respectively. As all green lines lie very close to the latter scenario, we expect the exponent for all 1:10 mergers to be similar and well above the observed relation (black solid line). All these models are a viable mechanism for size evolution even in more realistic scenarios, where dissipational eects would reduce the size growth (Covington et al., 2011; Hopkins et al., 2008).

7.4 Evolution of Surface Density

Next we take a closer look at the surface densities of the merger remnants. In the first and third panel of the left column in Fig. 7.3, the surface density of the equal-mass mergers grows at all radii, i.e. the lines are shifted more or less parallel to higher densities. This picture is the same for both major merger histories of one- (first left) and two- (third left) component models. Correspondingly, the mass assembles at all radii, which is depicted in the small panels beyond the respective surface densities. This evolution scenario is contrary to the observations of van Dokkum et al. (2010) (Fig. 2.2, Chapter 2), which show, that the compact early-type galaxies grow inside-out, i.e. the central densities stay constant and most of the mass assembles at larger radii, building up an extended envelope of stars.

The second column depicts the surface densities and mass assembly of the minor mergers with an initial mass ratio of 1:5. For the bulge only models (top) with a diuse satellite (Sat 1:5 in Fig. 7.1), the surface density stays nearly constant out to a radius of $r \approx 1 - 2$kpc and increases mainly in the outer parts. This behavior can also be seen for the corresponding mass assembly. The solid lines in the last two panels of the second column show that the same scenario is even more e cient using two-component models. Due to the massive dark matter halo, most of the bulge particles get stripped at larger radii and the central surface density stays unaected. Therefore, it just increases at radii $r > 2 - 3$kpc and most of the size growth is due to the build up of a massive stellar envelope. The dotted line in these panels depict the four remnants, where the satellites are more compact (cSat 1:5 in Fig. 7.1) and lie on the $z \sim 2$

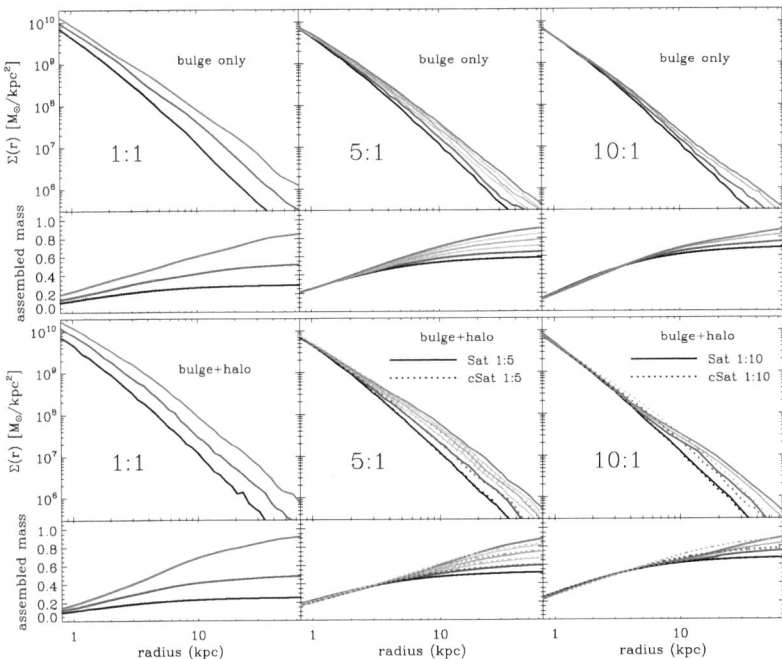

Figure 7.3: First row of panels: Surface densities for the bulge only models. For the equal-mass mergers (left), they increase at all radii and for five generations (from black to red) of minor mergers with an initial mass ratio of 1:5 (middle) they grow more in the outer regions. For an initial mass ratio of 1:10 (right), we can see the same behavior, i.e. after the second (blue), forth (green) and sixth (red) generation the surface density slightly increases at large radii and stays constant in the center. The panels of the second row show the mass assembly according to the surface densities of the top panels. Third row: Surface densities of the corresponding two-component models. Again, for equal-mass mergers (left), the surface density gets shifted parallel to higher values, but for a higher initial mass ratio of 1:5 (1:10), we can clearly see, that the central surface densities stay constant and a lot of particles assemble at radii larger than $r > 2$kpc ($r > 4$kpc), which is very similar to the inside-out growth scenario of van Dokkum et al. (2010) (see also Fig. 2.2 in Chapter 2). Regarding the according mass assembly of the bulge+halo models (bottom row), it is even more obvious, that the galaxies grow inside-out. The dotted lines for the 1:5 minor mergers indicate, that the accretion of a more compact satellite (cSat 1:5) yields the same results. But in the 1:10 scenario with compact satellites, more material goes further towards the center and less material assembles at large radii. Nevertheless, the central density also stays constant (for $r < 2$) and most of the accreted particles build up an extended envelope.

relation of Williams et al. (2010). Obviously, as the scale radii are very similar, the results stay the same.

The six generations of minor mergers with an initial mass ratio of 1:10 are shown in the last column of Fig. 7.3. In the case of bulge only models (top), the surface density grows predominantly at larger radii, similar to the previous scenario, but now the satellite is even less bound compared to the 1:5 case and therefore, it gets destroyed rapidly, even without a dark matter halo. In the case of two-component minor mergers of diuse satellites (solid lines, third and f orth panel) this eect becomes enhanced, as the satellite first orbits through the massive dark matter halo, before it gets closer to the host's center. Then all the material gets stripped at very large radii and the surface density stays constant out to a radius of $r = 5$kpc. Regarding the mass assembly, this is even more obvious, as the central mass stays constant out to a radius of $r \sim 10$kpc. Therefore, this scenario seems to be very e cient, as the outer surface density increases significantly, although the total amount of added mass is 40% less than for the 1:5 hierarchy, where the initial host mass gets doubled. However, this evolution scenario might be too extreme compared to observations (Fig. 2.2) and we can rule out the very diuse satellites at a redshift of $z \sim 2$.

This picture changes, if we use the more bound, compact satellite (cSat 1:10), depicted with the dotted lines in the last two panels. As the scale length of this satellite is two times smaller, it is much more bound and resists the drag force of the host potential for a longer time. Consequently, more material gets closer to the central regions, the remnant's surface densities grow outside a radius of $r > 2$kpc and more mass assembles at smaller radii. Regarding the according mass assembly (last panel), it grows predominantly outside a radius of 5kpc, which is also more consistent with the observed evolution.

In Fig. 7.4 we show the evolution of a full set (10 generations) of two-component minor mergers with a compact satellite (cSat 1:10, Fig. 7.1), but due to much lower computation time, we took radial orbits. Comparing the surface densities of this sequence (solid lines, Fig. 7.4) with the four generations with angular momentum (dotted lines, Fig. 7.3 and 7.4), they evolve nearly the same. The surface densities (top panel) stay constant out to a radius of $r \approx 2$kpc and the high size growth of a factor of ≈ 4.5 (see section 7.3) is driven by building up an extended envelope of luminous material. The mass assembly also looks very promising, as most of the particles accrete at radii larger than 5kpc. The results of this scenario are very similar to the 1:5 minor mergers of two-component models, which nicely resemble the observations (van Dokkum et al., 2010).

Altogether that means, that the mass assembly in minor mergers strongly depends on the eect of dynamical friction and tida l stripping, which of course are much more e cient for the two-component models, whe re the dark matter strips the particles of the in-falling satellites at exactly the right regions of the initial host galaxy. As consequence, nearly no material accretes in the central regions and therefore the central surface density stays constant. Looking at the accretion of diuse satellites in the 1:10 scenario, they seem to be too extreme and lose their material at too large radii. How-

7.4 Evolution of Surface Density

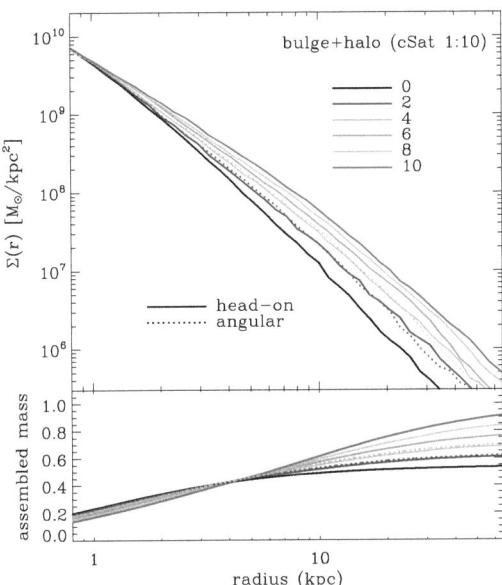

Figure 7.4: Top panel: Surface densities along the major axis for 10 generations of head-on minor mergers with the compact bulge+halo satellite cSat 1:10 (Fig. 7.1). The di erent colors give the generation and the dotted lines highlight the compact minor mergers with angular momentum of the last panel in Fig. 7.3. Bottom panel: Assembled mass plotted against the radius, as in Fig. 7.3. As our pericentric distances are very small, both scenarios show a very similar evolution, i.e. the more compact satellites go further to the center, compared to the less bound satellites (Sat 1:10, Fig. 7.1), and the host assembles mass at radii $r > 5$kpc.

ever, for the other bulge+halo models our results are in good agreement to observations, where the compact early-type galaxies make up the central cores of today's elliptical (Hopkins et al., 2009a; Bezanson et al., 2009; van Dokkum et al., 2010; Szomoru et al., 2011). Without a dark matter halo, the eect of dynamical friction is much weaker and the minor mergers of bulge only systems give less promising results.

7.5 Profile Shape Evolution

For a long period, the surface brightness profiles of elliptical galaxies have been fitted by the de Vaucouleurs $r^{1/4}$ profile (de Vaucouleurs, 1948). But more recent work shows, that the curvature of the light profiles seems to be very important as it correlates to other observed properties of elliptical galaxies, such as the eective radius r_e, the total luminosity and the stellar mass (Caon et al., 1993; Nipoti et al., 2003; Naab & Trujillo, 2006; Kormendy et al., 2009). Therefore we use the Sersic $r^{1/n}$ (Sersic, 1968) profile to fit the surface brightness profiles of our simulations. The formula can be written as

$$I(r) = I_e \cdot 10^{-b_n((r/r_e)^{1/n}-1)}, \quad (7.1)$$

where the three free parameters are half of the total luminosity I_e, the eective radius r_e and the so called Sersic index n, which gives the shape of the profile. The factor b_n, which only depends on n, is chosen such that the eective radius r_e encloses half of the total luminosity. For the expected range of Sersic indices, this factor can be approximated by the relation $b_n = 0.868n - 0.142$ (Caon et al., 1993). In the case of $n = 4$, Eq. 7.1 reduces to the de Vaucouleurs $r^{1/4}$ law.

In order to get a better comparison to recent observations of elliptical galaxies (e.g. Trujillo et al. 2004; Kormendy et al. 2009) we convert the projected surface densities of section 7.4 to a V-band surface brightness. Assuming a constant mass-to-light ratio, the radial luminosity profile can be written as

$$L(r) = (\ r) \cdot 10^{-M_V/2.5}, \quad (7.2)$$

where $M_V = 7.1973$ is the absolute magnitude of a star in the V-band at a distance of $10pc$, a stellar age of 10^{10}yr and close to solar metallicity $Z = 0.02$ (see Bruzual & Charlot 2003). ($\ r$) is the projected surface density of the previous section and the V-band magnitude can be calculated,

$$\mu_V(r) = -2.5 \cdot \log L(r) + 21.5721, \quad (7.3)$$

where 21.5721 is just a factor to convert surface density to mag/arcsec^2. As we want to fit μ_V with a Sersic function we have to take the logarithm of Eq. 7.1 which yields

$$\mu(r) = -2.5 \log I(r) = \mu_e + c_n[(r/r_e)^{1/n} - 1], \quad (7.4)$$

with $c_n = 2.5 \cdot b_n$ and $\mu_e = -2.5 \log I_e$.

7.5 Profile Shape Evolution

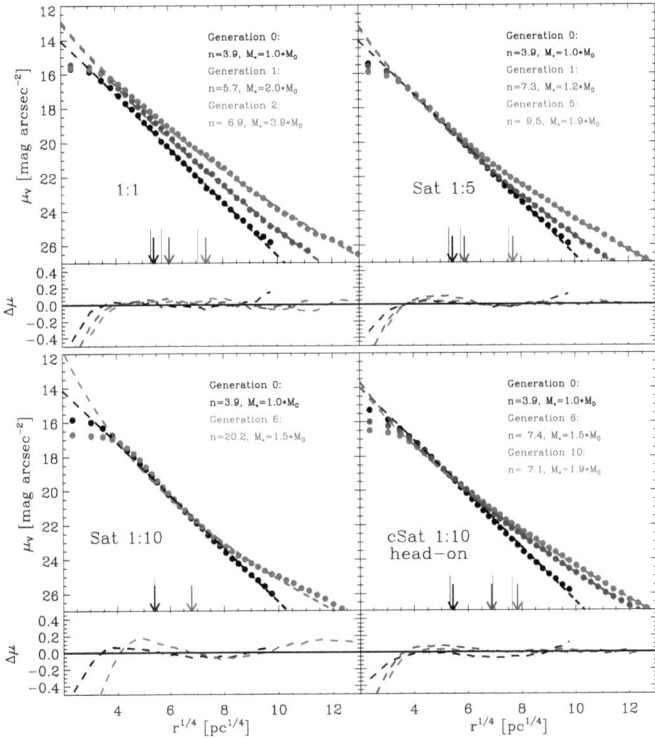

Figure 7.5: Top panels: Surface brightness profiles $\mu_V(r)$ of the two-component major (1:1) and minor merger (1:5) generations plotted against the radius. The black symbols depict the initial Hernquist profile, the blue symbols the profile of the first remnant and the red symbols the final profiles. The overplotted dashed lines in the corresponding colors show the best fitting Sersic function, which yields an increasing Sersic index n with each subsequent merger generation. The fitting range starts at $0.02 \cdot r_e$ (r_e is the eective radius of Fig. 7.2) and ends at either $10 \cdot r_e$ or a limiting surface brightness of $m_V = 27\mathrm{mag/arcsec}^2$, which results in residuals $\Delta\mu < 0.2\mathrm{mag/arcsec}^2$ (small panels below). As the profiles show an artificial core like structure (given by the initial conditions), the residuals increase for very small radii and the fitted eective radii $r_{e,fit}$ (thin lines at the bottom) are smaller than r_e of Fig. 7.1 (according arrows). Bottom panels: The same as in the four top panels, for the 1:10 minor merger with the diuse satellite (Sat 1:10, left) and for the full set of 10 generations with a compact satellite (cSat 1:10) and head-on orbits (right). In the first case, the final surface brightness profile shows a prominent kink, which results in an unrealistic high Sersic index $n \sim 20$ for the best fit. The minor mergers of compact satellites, show more reasonable results, but the Sersic index saturates rapidly at $n \approx 7 - 8$, which is due to the small amount of added mass for the final generations.

Figure 7.5 shows the Sersic fits to the surface brightness profiles of the equal-mass mergers (1:1, top left), the 1:5 (top right) and 1:10 minor mergers with angular momentum (bottom left) and the head-on scenario of Fig. 7.4 (bottom right). We chose a fitting range, so that we get a good fit to the main parts of the profile (> 95% along the major axis). If we start at $0.02 \cdot r_e$ and either go out to more than $10 \cdot r_e$ or to a limiting surface brightness of $m_V = 27$ mag/arcsec^{-2}, which is the limit of recent observations (Trujillo et al., 2004; Kormendy et al., 2009), the residuals are very small ($\Delta \mu < 0.2$mag/arcsec^2), except in the innermost regions, where the profiles have a core like structure. Looking at the profiles of the initial Hernquist spheres (black circles in all panels) we can see that we get a shape parameter of $n = 3.9$, which almost resembles the de Vaucouleurs profile ($n = 4$). As expected, the Hernquist sphere is a very good approximation of the de Vaucouleurs $r^{1/4}$ law over a large radial range and has a core profile in the innermost region (see also Naab & Trujillo 2006). Due to the core, the fitted Sersic profile overestimates the central surface brightness which leads to a fitted eective radius $r_{e,fit}$ (narrow vertical lines at the bottom of each surface brightness panel), which is slightly smaller compared to the 'real' eective radius r_e (corresponding arrows) of Fig. 7.1. This amount of 'artificial extra-light' from the fitted profile accounts for the discrepancy between these two radii, for all shown merger scenarios.

In the case of 1:1 mergers of two-component models (top left panel, Fig 7.5), we can see that the profile shape barely changes for the remnants. Therefore, the Sersic index shows only a small increase (see also black solid line, Fig. 7.6) compared to the added stellar mass M_*, which is not enough to explain the very high numbers, observers find for large core elliptical galaxies ($n \sim 10$, see Caon et al. 1993; Kormendy et al. 2009). This is a consequence of the violent merging process, where the material assembles at all radii and the surface brightness gets shifted nearly parallel to higher values, but does not significantly change the slope of the profile.

This picture changes dramatically if we go to higher initial mass ratios, where the merging process becomes dierent. In minor mergers violent relaxation does not aect the host galaxy (see Chapter 6), just the in-falling satellites. The latter one instantaneous feels the deep potential well of the host at closest approach and suers strongly from rapid potential fluctuations. Furthermore dynamical friction and tidal stripping get more prominent, as the satellites are more loosely bound than the host galaxy and the tidal forces are strong enough to strip a big amount of the satellite's material (see also Section 5.2).

Therefore, in the case of minor mergers, most of the accreted material assembles at larger radii of the galaxy and the central regions are hardly aected (see also section 7.4). Regarding the surface brightnesses of the 1:5 minor mergers (top right panel, Fig. 7.5), this implies, that the curvature, measured by the Sersic index n, changes rapidly with each further generation (see also red solid line, Fig. 7.6). Already after the first generation with a mass increase of a factor of ~ 1.2 we get a Sersic index of $n > 7$ and the final remnant has a slope of $n = 9.5$, which perfectly lies in the range of observations (Caon et al., 1993; Kormendy et al., 2009). The corresponding bulge only

7.5 PROFILE SHAPE EVOLUTION

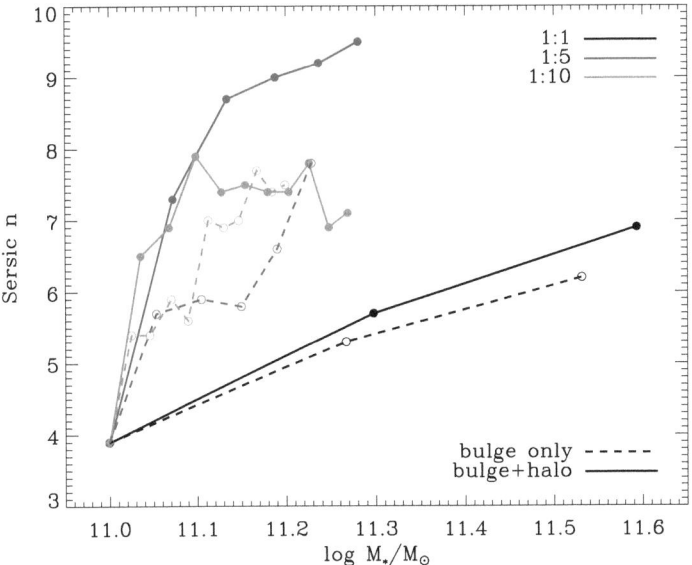

Figure 7.6: Evolution of the Sersic indices for all merger generations of Fig. 7.3, except the 1:10 scenario, which yields unrealistic fits (see bottom left, Fig. 7.5). As equal-mass mergers do not significantly change the slopes of the surface brightness the Sersic index after one generation is $n \sim 5-6$ (black lines). For the bulge+halo minor mergers with a mass ratio of 1:5 (red solid line), the slope increases rapidly for the first two generations before it converges to a final value of $n \sim 9.5$. The 1:10 head-on minor mergers with a compact satellite (green solid line) show the same trend, i.e. an initial fast increase of n for the first generations before it converges to a value of $n \sim 7-8$. For completeness, the dashed lines show the bulge only simulations, where the curvature for the minor mergers stays below $n = 8$.

scenario (red dashed line, Fig. 7.6) shows the same trend, and yields a final Sersic index of $n = 7.8$, but the overall evolution is much more e cient with two-component models. This again indicates, that the mass ratio and dark matter halo are very important, as they increase the eect of dynamical fri ction and tidal stripping in a way, that the accreted stellar mass assembles at the 'right' regions of the host galaxy, and leads to the observed profile shapes of core ellipticals.

If we further increase the initial mass ratio to 1:10 (bottom panels, Fig. 7.5), we can see that the Sersic index gets unrealistic large ($n > 20$) for the scenario with the diuse satellite (left panel). As the satellite is only weakly bound, it looses all its mass at very large radii, develops a kink at a radius of $r \approx$ 4kpc and the best fitting Sersic profile yields a very high curvature. This picture improves, if we take more bound satellites (cSat 1:10). Then the mass assembles more smoothly outside a radius of $r \approx$ 1.5kpc but still produces an extended outer envelope (right panel, Fig. 7.5). Although the evolution of the profiles look very reasonable, the Sersic index converges at a value of $n \approx 7 - 8$ (solid green line, Fig. 7.6), which can be explained with the very high mass ratio of the final generations (the last generation has a mass ratio of 1:19).

Altogether we can say that a massive dark matter halo enhances the eect of dynamical friction and tidal stripping. Considering satellites, which are not too weakly bound, it is the main driver to accrete the luminous matter at the 'right' regions. Then we also get reasonable results for the evolution of the Sersic index of $n \sim 8 - 10$ (Fig. 7.6). In the case of equal-mass mergers the eect of violent relaxation and mixing is more dominant and does not change the slope of the surface brightness profiles and we only get a mild increase after one generation $n \sim 5 - 6$ (Fig. 7.6).

7.6 Dark Matter Fractions

In this section we compare the dark matter fractions of our simulations with recent lensing observations, which predict an increasing dark matter fraction for more massive early-type ellipticals (Barnabè et al., 2011). In Fig. 7.7 we illustrate the dark matter fractions f_{dm} for all bulge+halo simulations

$$f_{dm}(r < r_{50}) = M_{dm}(r < r_{50})/M_{tot}(r < r_{50}), \qquad (7.5)$$

where r_{50} denotes the spherical half-mass radius of the stellar component and M_{dm}, M_{tot} are the halo and total masses within r_{50}. Obviously, the dark matter fraction increases rapidly with each subsequent minor merger generation regardless of which mass ratio. This strong evolution with additional mass is a consequence of the rapid size growth (Fig. 7.2, see also Section 6.5.2), which is in good agreement with (Nipoti et al., 2009b), who finds similar results in his numerical simulations. As the evolution of f_{dm} strongly correlates with the final radii of the merger remnants, the 1:5 scenario with a more compact satellite (red dashed line), which indicates the weakest size growth (Fig. 7.2), also has the lowest dark matter fractions of all minor merger scenarios. On the other hand, all 1:10 bulge+halo scenarios grow rapidly with mass and therefore have the highest and comparable evolutions for the dark matter fraction.

7.6 Dark Matter Fractions

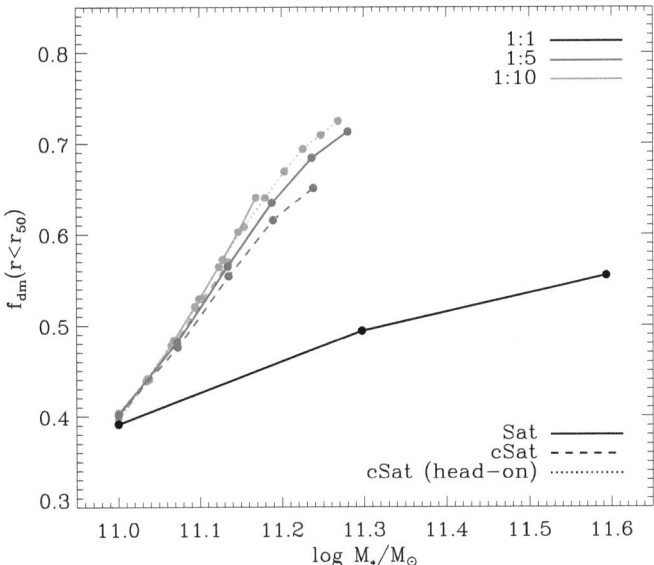

Figure 7.7: Evolution of the dark matter fraction f_{dm} within the spherical half-mass radius r_{50} for all bulge+halo mergers of this chapter. Due to the highly increasing radii (Fig. 7.2), the dark matter fractions grow significantly for all minor merger scenarios. As the size increase of the 1:5 scenario with a compact satellite (red dashed line) is the least e cient of, its dark matter fraction illustra tes the lowest growth. For all 1:10 scenarios, f_{dm} evolves very similar, as their sizes all grow by nearly the same amount (Fig. 7.2). The dark matter fraction of the equal-mass mergers (black line) indicate only a marginal evolution, which is due to the low size growth. However, as we have seen in Section 6.4.1, major mergers really change the dark matter fraction, as the violent merging process mixes more dark matter than stellar particles into the center.

In the case of equal-mass mergers, this looks somewhat different, i.e. they only show a very small increase of the dark matter fraction, compared to the added mass (black line, Fig. 7.7). But as we already have seen in Section 6.4.1, the merging process changes completely and the approaching galaxies suffer much more from violent relaxation and mixing. Consequently, relatively more dark matter than stellar particles get scattered to the central regions, which finally yields a real increase of the dark matter fraction. This result is contrary to (Nipoti et al., 2009b), who argues that regardless of the merging process, the dark matter fraction increases just due to the size growth.

Going back to the observations of Barnabè et al. (2011), we can see, that our simplified minor merger hierarchies can reproduce the evolution of the dark matter fractions of early-type ellipticals. Furthermore, our results are in quantitative good agreement to their results using a Chabrier IMF (Chabrier, 2003). But as our simulations are scale free, we can easily rescale the mass range of our remnants, thus they better fit the results using a Salpeter IMF (Salpeter, 1955).

7.7 Discussion and Conclusion

We have performed a set of numerical merger simulations of galaxy models which consist of either a one-component (bulge only) or a two-component (bulge+ dark matter halo) Hernquist profile. Furthermore we used different initial mass ratios and orbits with and without angular momentum to cover a large range of parameters. Our main findings can be summarized as follows. Dry equal-mass mergers can not be the main driver of galaxy evolution, because the remnants assemble mass at all radii and shift the surface density nearly parallel to higher values. Therefore the slope of the surface brightness profiles are hardly affected and the Sersic index does not exceed a value of $n \sim 6$ after one generation (Fig. 7.6). Additionally the size growth per added mass is limited to $r_e \propto M^{0.91}$.

On the other hand, dissipationless minor mergers show very promising results. As the satellites are loosely bound, they strongly suffer from dynamical friction and tidal stripping. Therefore, depending on the mas ratio, most of the mass assembles at larger radii and the remnants develop a extended envelope of stars. For an initial mass ratio of 1:5, the central surface density/brightness stays constant, out to a radius of $r \approx 1.5$ kpc which is in good agreement to recent observations of high redshift early-types (Szomoru et al., 2011) and the assumption, that the compact galaxies are the cores of present day ellipticals (Hopkins et al., 2009a; Bezanson et al., 2009; van Dokkum et al., 2010; Szomoru et al., 2011). Furthermore the fitted Sersic profiles yield a curvature of $n \sim 8-10$, which lies in the range of most of the observed core ellipticals (Caon et al., 1993; Kormendy et al., 2009). For a higher initial mass ratio, the results highly depend on the properties of the satellite galaxies. If the satellites are loosely bound (Sat 1:10, Fig. 7.1), they are destroyed rapidly in the case of more realistic bulge+halo models and the host's central surface density stays constant out

7.7 Discussion and Conclusion

to a radius of $r \approx 5\text{kpc}$. Additionally, the developing envelope of accreted particles reveals a prominent kink, which results in an unrealistic high Sersic index of $n \approx 20$. Using a compact, more bound, satellite (cSat 1:10, Fig. 7.1), the accreted mass gets closer to the center and distributes more smoothly in the outer envelope, which then yields reasonable Sersic indices of $n \approx 7-8$.

Regarding the size growth of the individual minor merger scenarios, we find that all remnants, where the bulge is embedded in a massive dark matter halo, lie above the grey shaded area of Fig. 7.2 and are viable drivers for the evolution of compact early-type ellipticals. In the most promising cases, namely the 1:5 sequence with a diuse satellite and the 1:10 sequence wit h a compact satellite result in a mass-size relation of $r_e \propto M^{2.3}$ and $r_e \propto M^{2.5}$, respectively. On the other hand, for the bulge only minor mergers, only the hierarchy with a mass ratio of 1:10 and a diuse, weakly bound satellite evolves in excess of the observed mass-size relation. The other scenarios stay within the 'forbidden' area and are by far not e cient enough to give a proper size increase.

Furthermore, in Section 6, we have shown, that the merging history of our galaxy models yield dark matter fractions, which are in good agreement to recent lensing observations of the SLACS collaboration. Our minor merger remnants also indicate, that the central dark matter fractions increase with the stellar mass of early-type ellipticals.

Combining all the results we can say, that only the minor mergers including a dark matter halo give reasonable results for the observed inside-out growth of the surface densities and surface brightnesses. Furthermore, only the bulge+halo minor mergers always yield a size growth in excess to the observed prediction of van Dokkum et al. (2010) (see Fig. 2.2).

However, we use a very idealized scenario, without any gas or black hole physics. But even the existence of a small amount of gas, which is known to reduce the size growth (Covington et al., 2011; Hopkins et al., 2008), might not be e cient enough, as we achieve very high growth rates in most of the minor merger scenarios. On the other hand, the implementation of black holes might boost the 'pu up' scenario, as black hole binaries are able to deplete the central galaxy regions from gas and stars (Fan et al., 2008, 2010), and the observed core structure of the most massive present ellipticals will get more prominent.

CHAPTER 8
CONCLUSION & OUTLOOK

Recent observations have revealed a population of compact high redshift ($z \sim 2-3$) early-type galaxies, which are a factor 4-5 smaller than their present day counterparts. They are already very massive but have less concentrated surface density profiles, represented by small Sersic indices of $n \sim 4$. Furthermore, the stellar populations of their present day counterparts indicate, that dissipation and star formation cannot be the main evolutionary mechanism. However, in the currently favoured cosmological model, where structure grows hierarchically, dissipationless mergers are supposed to be the main driver for the subsequent evolution of compact, high redshift early-type galaxies. Therefore we employ a large set of more than 80 dissipationless merger simulations, with a large avriety of orbits and initial agalxy models.

To achieve this, we first created an initial condition program, which covers a wide range of dierent parameters (Chapter 4). In detail, this means, that we can either chose galaxies consisting of just a stellar component or a more realistic two-component model, where the stellar bulge is embedded in a more massive dark matter halo. The dark matter halo is fixed to a Hernquist density distribution (Hernquist, 1990), but the slope of the luminous part can vary between dierent -models (Dehnen, 1993; Tremaine et al., 1994). Hence, one can adopt either a steep density slope ~ 2, resembling a cuspy extra-light galaxy, or a cored profile $= 0$, which is observed for the most massive ellipticals. First stability tests have shown, that the initial conditions are very stable for one- and two-component galaxy models with dierent particle resolutions and density slopes but equal mass particles. We also created more 'realistic' galaxy models using the most recent HOD models to get a viable dark to stellar mass ratio of $M_*/M_{dm} \sim 0.015$ (Moster et al., 2010; Wake et al., 2011) for a $M_* = 10^{11} M_\odot$ galaxy at a redshift of $z \sim 2$. In addition, we assigned them a size corresponding to the most recent mass-size relations of early-type ellipticals (Williams et al., 2010) at this high redshift. In this setup, the dark matter particles have a much higher mass than the stellar particles, and we have to use a much larger softening length compared to the equal-mass models. This stems from two-body relaxation in the central high density regions, which additionally induces mass segregation for unequal mass particles (Chapter 5). Therefore, with a larger force softening length, we prevent the more massive dark matter particles to sink to the center and kick out the less massive stellar particles. The adopted softening length is still small compared to the eective galaxy radius, so that the galaxies can be considered resolved and we obtain very stable results for cosmologically motivated galaxy models.

Using our well tested initial conditions, we are able to investigate the dynamics of the merging process with an unprecedented high accuracy and resolution. First we focused on equal-mass mergers of either one- or two-component models and found, that violent relaxation and mixing are the dominant processes, which significantly change the structure of the merger remnant's dierential energy distribution (Section 6.4). Strong potential fluctuations during the closest encounters oer new energy states with higher binding energies and unbind a non-negligible amount of initial weakly bound particles (see also White 1978). This evolution is the same for one- and two-component models, but in the latter case we display another striking result in the galaxy's center, where the amount of dark matter particles increases with respect to the stellar particles and we obtain a 'real' increase of the central dark matter fraction. This is contrary to the work of Nipoti et al. (2009b), who argues, that the increase of the dark matter fraction is just an eect of the increasing galaxy size. However, we convincingly show, that violent relaxation mixes more and more dark matter particles to higher binding energies for each subsequent merger generation, which changes the ratio of stellar to dark matter particles, especially in the central regions. Furthermore, the redistribution of the particle's energies causes the systems to seek for a new equilibrium configuration. Therefore, the final velocity dispersion profiles of the equal-mass merger remnants can nicely be fitted by a Jae profile (Jae , 1983), which is in good agreement to Spergel & Hernquist (1992). Unfortunately, the structural changes and a transfer of energy from the bulge to the halo increase the velocities of the merger remnants and consequently limit the size growth. However, the mass-size ($r_e \propto M_*^{0.8-1.0}$) and the mass-velocity relation ($M_* \propto {}_e^{3.3-5.1}$) yields reasonable results compared to a previous study of Boylan-Kolchin et al. (2005).

Looking at the minor merger scenarios with initial mass ratios of 1:10, we worked out, that first of all, the dynamics is completely dierent and the host galaxy is not aected by violent relaxation (Section 6.5). Therefore its central properties are conserved and the velocity dispersion profiles of the total systems stay constant for both, one- and two-component models. But as the satellites are loosely bound compared to the host galaxies, they strongly suer from tidal stripping and nearly no material reaches the host's center. The stripped material builds up an extended envelope of stars, which significantly boosts the size growth. Due to an enhanced eect of dynamical friction and tidal stripping in the dark matter halo of the bulge+halo model, this mass assembly is very e cient, and we get a final size increase of more than a factor of ~ 4.5, which is exactly in the observed range for the size growth of compact high redshift ellipticals (Szomoru et al., 2011). Looking at the mass-size relations of all two-component scenarios with mass ratios of 1:5 and 1:10, we get $r_e \propto M_*^{>2.3}$, which is much steeper than the result of van Dokkum et al. (2010) ($r_e \propto M_*^2$).

As minor mergers seem to be very good candidates for driving the evolution of early-type ellipticals we extend the previous set of simulations by a sequence with an initial mass ratio of 1:5, investigate in more detail the assembly history of the final remnants, and want to figure out the importance of a dark matter halo (Chapter 7). First of all, regarding the size evolution of the individual scenarios, we can see that only the minor mergers with dark matter halo evolve as expected from observations. For bulge only models, just one merger configuration with a very diuse satellite galaxy evolves in excess to the observed relation. Looking at the evolution of the surface densities, recent observations reveal an inside-out growth with decreasing redshift (van Dokkum et al., 2010; Szomoru et al., 2011). In detail this implies, that the central regions of high redshift early types stay unaected and build the cores of present day ellipticals, while they assemble a lot of stellar mass at larger radii, building up an outer envelope. Surprisingly, this is exactly what we find for the surface densities of our minor merger remnants (Section 7.4). While the profiles of the equal-mass mergers grow over the whole radial range, the central profiles of the minor merger scenarios stay constant, as most of the satellite's material gets stripped in the outer regions of the host galaxy. Again, the scenarios with bulge+halo models yield more promising results, as the extended massive dark matter halo enhances tidal eects and strips the material at the 'right' regions ($r > 2$kpc). Converting the surface densities of our remnants to surface brightness profiles, Sersic fits indicate that for the first generations, the Sersic index n increases most rapidly for the two-component minor mergers of both mass ratios (1:5, 1:10). Although n converges to a value between $n = 7 - 8$ for the 1:10 sequence, the 1:5 sequence of bulge+halo models is the only scenario, where we get a high Sersic index of $n \sim 9 - 10$, which is expected from observations. Finally, we obtain an increasing dark matter fraction, which is consistent with recent observations (Barnabè et al., 2011) for all minor merger hierarchies and we can conclude, that the existence of a dark matter halo is not just expected but is essential to get a viable evolution scenario.

Altogether we can say, that we highlight a very promising scenario to close the gap between the compact high redshift ellipticals and their more extended counterparts in the present day Universe. To fortify our results, the next step would be to use more realistic galaxy models, applying a contracted NFW-profile for the halo and extend the galaxy by the potential of a supermassive black hole. Fan et al. (2008) showed, that AGNs can also pu up galaxies and might even improve the results of the dissipationless merger scenario. Regarding the dynamics of merging systems, it would be desireable to investigate the impact of dierent orbital properties, to estimate the eect of, e.g. hyperbolic or bound orbits with dierent im pact parameters. Furthermore, we started to look at the evolution of the very tight relation between a galaxy's escape velocity v_{esc} and its metallicity (Scott et al., 2009).

Therefore we simply compute a particle's escape velocity of the initial host and satellite galaxy and assign to it the according metallicity. During each merger genera-

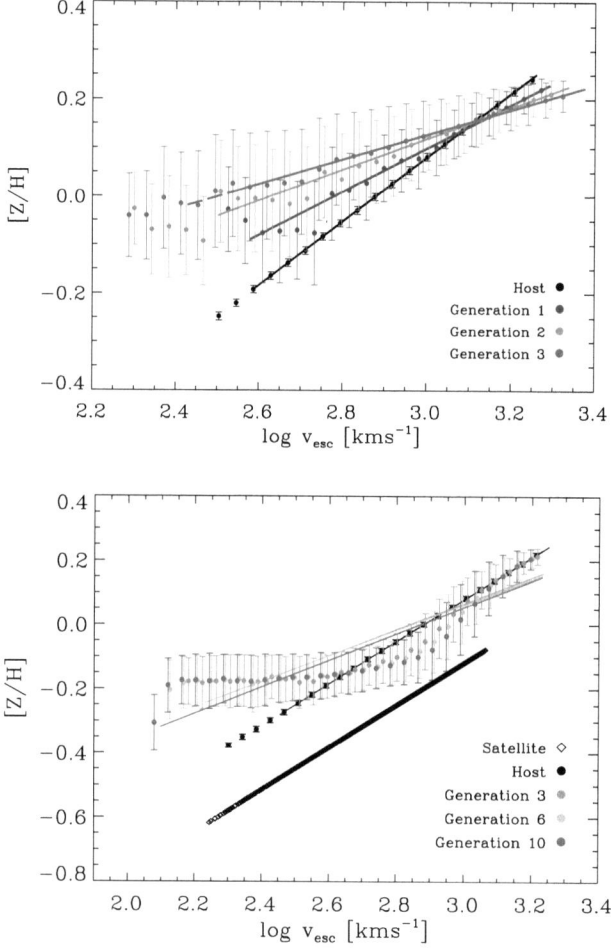

Figure 8.1: Top panel: Evolution of the v_{esc}-metallicty relation for three generations of equal-mass mergers. Due to the strong mixing, it changes at all velocities, and introduces an increasing scatter for each subsequent generation. The solid lines are the best linear fits to the corresponding data points. Bottom panel: Same as above for 10 generations of 1:10 minor mergers. Obviously, the central relation does not change and only the outerparts with lower velocities are eect. Therefore the metallicity seems to converge to a value of $[Z/H] = -0.2$. Furthermore, the scatter is much smaller and the relation stays very tight, compared to the equal-mass mergers.

tion, the escape velocities of the particles change, but their metallicities stay constant, thus we can evaluate the merger induced scatter in the v_{esc}- metallicity relation. As observations show a very tight correlation, we can approximate the contribution of minor or major mergers for the evolution of elliptical galaxies. The first results indicate, that the $v_{esc} - [Z/H]$-relation for equal-mass mergers (top panel, Fig. 8.1) changes at all radii and for all velocities, which stems from the strong mixing, induced by violent relaxation. Therefore, the scatter increases significantly and the overall relation becomes very broad. On the other hand, a sequence of ten minor mergers with initial mass ratio of 1:10 (bottom panel, Fig. 8.1) introduces only a small scatter in the $v_{esc} - [Z/H]$-relation and has no influence on the central regions (with high velocities). Furthermore, with each generation, the metallicity gradient becomes only weaker for equal-mass mergers (Fig. 8.1), which is in good agreement with earlier predictions of White (1978) and Villumsen (1982). Further details of this scenario have to be tested.

BIBLIOGRAPHY

AARSETH, S. J.: 1963. Dynamical evolution of clusters of galaxies, I. *MNRAS*, **126**, 223.

ABRAMOWITZ, M. STEGUN, I. A.: 1970. *Handbook of mathematical functions : with formulas, graphs, and mathematical tables*.

ARAD, I. LYNDEN-BELL, D.: 2005. Inconsistency in theories of violent relaxation. *MNRAS*, **361**, 385–395.

AUGER, M. W., TREU, T., BOLTON, A. S., GAVAZZI, R., KOOPMANS, L. V. E., MARSHALL, P. J., BUNDY, K., MOUSTAKAS, L. A.: 2009. The Sloan Lens ACS Survey. IX. Colors, Lensing, and Stellar Masses of Early-Type Galaxies. *ApJ*, **705**, 1099–1115.

AUGER, M. W., TREU, T., BOLTON, A. S., GAVAZZI, R., KOOPMANS, L. V. E., MARSHALL, P. J., MOUSTAKAS, L. A., BURLES, S.: 2010. The Sloan Lens ACS Survey. X. Stellar, Dynamical, and Total Mass Correlations of Massive Early-type Galaxies. *ApJ*, **724**, 511–525.

BAES, M., DEJONGHE, H., BUYLE, P.: 2005. The dynamical structure of isotropic spherical galaxies with a central black hole. *A&A*, **432**, 411–422.

BARNABÈ, M., CZOSKE, O., KOOPMANS, L. V. E., TREU, T., BOLTON, A. S.: 2011. Two-dimensional kinematics of SLACS lenses - III. Mass structure and dynamics of early-type lens galaxies beyond z 0.1. *MNRAS*, **415**, 2215–2232.

BARNES, J. HUT, P.: 1986. A hierarchical O(N log N) force-calculation algorithm. *Nature*, **324**, 446–449.

BARNES, J. E.: 1992. Transformations of galaxies. I - Mergers of equal-mass stellar disks. *ApJ*, **393**, 484–507.

BEHROOZI, P. S., CONROY, C., WECHSLER, R. H.: 2010. A Comprehensive Analysis of Uncertainties Aecting the Ste llar Mass-Halo Mass Relation for $0 < z < 4$. *ApJ*, **717**, 379–403.

BELL, E. F., NAAB, T., MCINTOSH, D. H., SOMERVILLE, R. S., CALDWELL, J. A. R., BARDEN, M., WOLF, C., RIX, H.-W., BECKWITH, S. V., BORCH, A., HÄUSSLER, B., HEYMANS, C., JAHNKE, K., JOGEE, S., KOPOSOV, S., MEISENHEIMER, K., PENG, C. Y., SANCHEZ, S. F., WISOTZKI, L.: 2006a. Dry Mergers in GEMS: The Dynamical Evolution of Massive Early-Type Galaxies. *ApJ*, **640**, 241–251.

BELL, E. F., PHLEPS, S., SOMERVILLE, R. S., WOLF, C., BORCH, A., MEISENHEIMER, K.: 2006b. The Merger Rate of Massive Galaxies. *ApJ*, **652**, 270–276.

BENDER, R., BURSTEIN, D., FABER, S. M.: 1992. Dynamically hot galaxies. I - Structural properties. *ApJ*, **399**, 462–477.

BENDER, R., BURSTEIN, D., FABER, S. M.: 1993. Dynamically hot galaxies. II - Global stellar populations. *ApJ*, **411**, 153–169.

BENDER, R., SURMA, P., DOEBEREINER, S., MOELLENHOFF, C., MADEJSKY, R.: 1989. Isophote shapes of elliptical galaxies. II - Correlations with global optical, radio and X-ray properties. *A&A*, **217**, 35–43.

BERNARDI, M.: 2009. Evolution in the structural properties of early-type brightest cluster galaxies at small lookback time and dependence on the environment. *MNRAS*, **395**, 1491–1506.

BEZANSON, R., VAN DOKKUM, P. G., TAL, T., MARCHESINI, D., KRIEK, M., FRANX, M., COPPI, P.: 2009. The Relation Between Compact, Quiescent Highredshift Galaxies and Massive Nearby Elliptical Galaxies: Evidence for Hierarchical, Inside-Out Growth. *ApJ*, **697**, 1290–1298.

BINDONI, D. SECCO, L.: 2008. Violent relaxation in phase-space. , **52**, 1–18.

BINNEY, J. TREMAINE, S.: 2008. *Galactic Dynamics: Second Edition*. Princeton University Press.

BOLTON, A. S., BURLES, S., KOOPMANS, L. V. E., TREU, T., GAVAZZI, R., MOUSTAKAS, L. A., WAYTH, R., SCHLEGEL, D. J.: 2008. The Sloan Lens ACS Survey. V. The Full ACS Strong-Lens Sample. *ApJ*, **682**, 964–984.

BOURNAUD, F., CHAPON, D., TEYSSIER, R., POWELL, L. C., ELMEGREEN, B. G., ELMEGREEN, D. M., DUC, P.-A., CONTINI, T., EPINAT, B., SHAPIRO, K. L.: 2011. Hydrodynamics of High-redshift Galaxy Collisions: From Gas-rich Disks to Dispersion-dominated Mergers and Compact Spheroids. *ApJ*, **730**, 4.

BOYLAN-KOLCHIN, M. MA, C.-P.: 2004. Major mergers of galaxy haloes: cuspy or cored inner density profile? *MNRAS*, **349**, 1117–1129.

BOYLAN-KOLCHIN, M., MA, C.-P., QUATAERT, E.: 2005. Dissipationless mergers of elliptical galaxies and the evolution of the fundamental plane. *MNRAS*, **362**, 184–196.

BOYLAN-KOLCHIN, M., MA, C.-P., QUATAERT, E.: 2006. Red mergers and the assembly of massive elliptical galaxies: the fundamental plane and its projections. *MNRAS*, **369**, 1081–1089.

BOYLAN-KOLCHIN, M., MA, C.-P., QUATAERT, E.: 2008. Dynamical friction and galaxy merging time-scales. *MNRAS*, **383**, 93–101.

BOYLAN-KOLCHIN, M., SPRINGEL, V., WHITE, S. D. M., JENKINS, A., LEMSON, G.: 2009. Resolving cosmic structure formation with the Millennium-II Simulation. *MNRAS*, **398**, 1150–1164.

BRUZUAL, G. CHARLOT, S.: 2003. Stellar population synthesis at the resolution of 2003. *MNRAS*, **344**, 1000–1028.

BUITRAGO, F., TRUJILLO, I., CONSELICE, C. J., BOUWENS, R. J., DICKINSON, M., YAN, H.: 2008. Size Evolution of the Most Massive Galaxies at $1.7 < z < 3$ from GOODS NICMOS Survey Imaging. *ApJ*, **687**, L61–L64.

BULLOCK, J. S., KOLATT, T. S., SIGAD, Y., SOMERVILLE, R. S., KRAVTSOV, A. V., KLYPIN, A. A., PRIMACK, J. R., DEKEL, A.: 2001. Profiles of dark haloes: evolution, scatter and environment. *MNRAS*, **321**, 559–575.

CAON, N., CAPACCIOLI, M., D'ONOFRIO, M.: 1993. On the Shape of the Light Profiles of Early Type Galaxies. *MNRAS*, **265**, 1013–+.

CAPELATO, H. V., DE CARVALHO, R. R., CARLBERG, R. G.: 1995. Mergers of Dissipationless Systems: Clues about the Fundamental Plane. *ApJ*, **451**, 525.

CAPPELLARI, M., DI SEREGO ALIGHIERI, S., CIMATTI, A., DADDI, E., RENZINI, A., KURK, J. D., CASSATA, P., DICKINSON, M., FRANCESCHINI, A., MIGNOLI, M., POZZETTI, L., RODIGHIERO, G., ROSATI, P., ZAMORANI, G.: 2009. Dynamical Masses of Early-Type Galaxies at $z \sim 2$: Are they Truly Superdense? *ApJ*, **704**, L34–L39.

CENARRO, A. J. TRUJILLO, I.: 2009. Mild Velocity Dispersion Evolution of Spheroid-Like Massive Galaxies Since $z \sim 2$. *ApJ*, **696**, L43–L47.

CHABRIER, G.: 2003. Galactic Stellar and Substellar Initial Mass Function. *PASP*, **115**, 763–795.

CHANDRASEKHAR, S.: 1942. *Principles of stellar dynamics*.

CHANDRASEKHAR, S.: 1943. Dynamical Friction. I. General Considerations: the Coecient of Dynamical Friction. *ApJ*, **97**, 255.

CIMATTI, A., CASSATA, P., POZZETTI, L., KURK, J., MIGNOLI, M., RENZINI, A., DADDI, E., BOLZONELLA, M., BRUSA, M., RODIGHIERO, G., DICKINSON, M., FRANCESCHINI, A., ZAMORANI, G., BERTA, S., ROSATI, P., HALLIDAY, C.: 2008. GMASS ultradeep spectroscopy of galaxies at z $\sim$ 2. II. Superdense passive galaxies: how did they form and evolve? *A&A*, **482**, 21–42.

COLE, S., LACEY, C. G., BAUGH, C. M., FRENK, C. S.: 2000. Hierarchical galaxy formation. *MNRAS*, **319**, 168–204.

COVINGTON, M. D., PRIMACK, J. R., PORTER, L. A., CROTON, D. J., SOMERVILLE, R. S., DEKEL, A.: 2011. The role of dissipation in the scaling relations of cosmological merger remnants. *MNRAS*, **415**, 3135–3152.

DADDI, E., RENZINI, A., PIRZKAL, N., CIMATTI, A., MALHOTRA, S., STIAVELLI, M., XU, C., PASQUALI, A., RHOADS, J. E., BRUSA, M., DI SEREGO ALIGHIERI, S., FERGUSON, H. C., KOEKEMOER, A. M., MOUSTAKAS, L. A., PANAGIA, N., WINDHORST, R. A.: 2005. Passively Evolving Early-Type Galaxies at $1.4 < \sim z < \sim 2.5$ in the Hubble Ultra Deep Field. *ApJ*, **626**, 680–697.

DAMJANOV, I., MCCARTHY, P. J., ABRAHAM, R. G., GLAZEBROOK, K., YAN, H., MENTUCH, E., LEBORGNE, D., SAVAGLIO, S., CRAMPTON, D., MUROWINSKI, R., JUNEAU, S., CARLBERG, R. G., JØRGENSEN, I., ROTH, K., CHEN, H.-W., MARZKE, R. O.: 2009. Red Nuggets at z $\sim$ 1.5: Compact Passive Galaxies and the Formation of the Kormendy Relation. *ApJ*, **695**, 101–115.

DAVIS, M., EFSTATHIOU, G., FRENK, C. S., WHITE, S. D. M.: 1985. The evolution of large-scale structure in a universe dominated by cold dark matter. *ApJ*, **292**, 371–394.

DE LUCIA, G., SPRINGEL, V., WHITE, S. D. M., CROTON, D., KAUFFMANN, G.: 2006. The formation history of elliptical galaxies. *MNRAS*, **366**, 499–509.

DE VAUCOULEURS, G.: 1948. Recherches sur les Nebuleuses Extragalactiques. *Annales d'Astrophysique*, **11**, 247–+.

DEHNEN, W.: 1993. A Family of Potential-Density Pairs for Spherical Galaxies and Bulges. *MNRAS*, **265**, 250–+.

DEHNEN, W.: 2001. Towards optimal softening in three-dimensional N-body codes - I. Minimizing the force error. *MNRAS*, **324**, 273–291.

DEKEL, A., SARI, R., CEVERINO, D.: 2009. Formation of Massive Galaxies at High Redshift: Cold Streams, Clumpy Disks, and Compact Spheroids. *ApJ*, **703**, 785–801.

DJORGOVSKI, S. DAVIS, M.: 1987. Fundamental properties of elliptical galaxies. *ApJ*, **313**, 59–68.

DJORGOVSKI, S., DE CARVALHO, R., HAN, M.-S.: 1988. The universality(?) of distance-indicator relations. In S. van den Bergh & C. J. Pritchet, editor, *The Extragalactic Distance Scale*, volume 4 of *Astronomical Society of the Pacific Conference Series*, pages 329–341.

D'ONOFRIO, M., CAPACCIOLI, M., CAON, N.: 1994. On the Shape of the Light Profiles of Early Type Galaxies - Part Two - the - Diagram. *MNRAS*, **271**, 523.

DRESSLER, A., LYNDEN-BELL, D., BURSTEIN, D., DAVIES, R. L., FABER, S. M., TERLEVICH, R., WEGNER, G.: 1987. Spectroscopy and photometry of elliptical galaxies. I - A new distance estimator. *ApJ*, **313**, 42–58.

DUFFY, A. R., SCHAYE, J., KAY, S. T., DALLA VECCHIA, C.: 2008. Dark matter halo concentrations in the Wilkinson Microwave Anisotropy Probe year 5 cosmology. *MNRAS*, **390**, L64–L68.

EDDINGTON, A. S.: 1916. The distribution of stars in globular clusters. *MNRAS*, **76**, 572–585.

FABER, S. M.: 1987. Book-Review - Nearly Normal Galaxies from the Planck Time to the Present. *Science*, **238**, 1155.

FABER, S. M. JACKSON, R. E.: 1976. Velocity dispersions and mass-to-light ratios for elliptical galaxies. *ApJ*, **204**, 668–683.

FAN, L., LAPI, A., BRESSAN, A., BERNARDI, M., DE ZOTTI, G., DANESE, L.: 2010. Cosmic Evolution of Size and Velocity Dispersion for Early-type Galaxies. *ApJ*, **718**, 1460–1475.

FAN, L., LAPI, A., DE ZOTTI, G., DANESE, L.: 2008. The Dramatic Size Evolution of Elliptical Galaxies and the Quasar Feedback. *ApJ*, **689**, L101–L104.

FAROUKI, R. T., SHAPIRO, S. L., DUNCAN, M. J.: 1983. Hierarchical merging and the structure of elliptical galaxies. *ApJ*, **265**, 597–605.

FERRARESE, L., CÔTÉ, P., JORDÁN, A., PENG, E. W., BLAKESLEE, J. P., PIATEK, S., MEI, S., MERRITT, D., MILOSAVLJEVIĆ, M., TONRY, J. L., WEST, M. J.: 2006. The ACS Virgo Cluster Survey. VI. Isophotal Analysis and the Structure of Early-Type Galaxies. *ApJS*, **164**, 334–434.

FRANX, M., VAN DOKKUM, P. G., SCHREIBER, N. M. F., WUYTS, S., LABBÉ, I., TOFT, S.: 2008. Structure and Star Formation in Galaxies out to z = 3: Evidence for Surface Density Dependent Evolution and Upsizing. *ApJ*, **688**, 770–788.

GAVAZZI, R., TREU, T., KOOPMANS, L. V. E., BOLTON, A. S., MOUSTAKAS, L. A., BURLES, S., MARSHALL, P. J.: 2008. The Sloan Lens ACS Survey. VI. Discovery and Analysis of a Double Einstein Ring. *ApJ*, **677**, 1046–1059.

GAVAZZI, R., TREU, T., RHODES, J. D., KOOPMANS, L. V. E., BOLTON, A. S., BURLES, S., MASSEY, R. J., MOUSTAKAS, L. A.: 2007. The Sloan Lens ACS Survey. IV. The Mass Density Profile of Early-Type Galaxies out to 100 Eective Radii. *ApJ*, **667**, 176–190.

GENEL, S., GENZEL, R., BOUCHÉ, N., STERNBERG, A., NAAB, T., SCHREIBER, N. M. F., SHAPIRO, K. L., TACCONI, L. J., LUTZ, D., CRESCI, G., BUSCHKAMP, P., DAVIES, R. I., HICKS, E. K. S.: 2008. Mergers and Mass Accretion Rates in Galaxy Assembly: The Millennium Simulation Compared to Observations of z $\sim$ 2 Galaxies. *ApJ*, **688**, 789–793.

GONZÁLEZ-GARCÍA, A. C. VAN ALBADA, T. S.: 2005. Encounters between spherical galaxies - II. Systems with a dark halo. *MNRAS*, **361**, 1043–1054.

GRAHAM, A. COLLESS, M.: 1997. Some eects of galaxy structure and dynamics on the Fundamental Plane. *MNRAS*, **287**, 221–239.

GRAHAM, A., LAUER, T. R., COLLESS, M., POSTMAN, M.: 1996. Brightest Cluster Galaxy Profile Shapes. *ApJ*, **465**, 534.

GRAHAM, A. W.: 2001. An Investigation into the Prominence of Spiral Galaxy Bulges. *AJ*, **121**, 820–840.

GRILLO, C. GOBAT, R.: 2010. On the initial mass function and tilt of the fundamental plane of massive early-type galaxies. *MNRAS*, **402**, L67–L71.

GUO, Q. WHITE, S. D. M.: 2008. Galaxy growth in the concordance CDM cosmology. *MNRAS*, **384**, 2–10.

GUO, Q. WHITE, S. D. M.: 2009. High-redshift galaxy populations and their descendants. *MNRAS*, **396**, 39–52.

HERNQUIST, L.: 1990. An analytical model for spherical galaxies and bulges. *ApJ*, **356**, 359–364.

HERNQUIST, L. KATZ, N.: 1989. TREESPH - A unification of SPH with the hierarchical tree method. *ApJS*, **70**, 419–446.

HOCKNEY, R. W. EASTWOOD, J. W.: 1981. *Computer Simulation Using Particles*.

HOPKINS, P. F., BUNDY, K., HERNQUIST, L., WUYTS, S., COX, T. J.: 2010. Discriminating between the physical processes that drive spheroid size evolution. *MNRAS*, **401**, 1099–1117.

HOPKINS, P. F., BUNDY, K., MURRAY, N., QUATAERT, E., LAUER, T. R., MA, C.: 2009a. Compact high-redshift galaxies are the cores of the most massive present-day spheroids. *MNRAS*, **398**, 898–910.

HOPKINS, P. F., COX, T. J., HERNQUIST, L.: 2008. Dissipation and the Fundamental Plane: Observational Tests. *ApJ*, **689**, 17–48.

HOPKINS, P. F., HERNQUIST, L., COX, T. J., KERES, D., WUYTS, S.: 2009b. Dissipation and Extra Light in Galactic Nuclei. IV. Evolution in the Scaling Relations of Spheroids. *ApJ*, **691**, 1424–1458.

JAFFE, W.: 1983. A simple model for the distribution of light in spherical galaxies. *MNRAS*, **202**, 995–999.

JOUNG, M. R., CEN, R., BRYAN, G. L.: 2009. Galaxy Size Problem at $z = 3$: Simulated Galaxies are too Small. *ApJ*, **692**, L1–L4.

KAZANTZIDIS, S., MAGORRIAN, J., MOORE, B.: 2004. Generating Equilibrium Dark Matter Halos: Inadequacies of the Local Maxwellian Approximation. *ApJ*, **601**, 37–46.

KEREŠ, D., KATZ, N., FARDAL, M., DAVÉ, R., WEINBERG, D. H.: 2009. Galaxies in a simulated CDM Universe - I. Cold mode and hot cores. *MNRAS*, **395**, 160–179.

KEREŠ, D., KATZ, N., WEINBERG, D. H., DAVÉ, R.: 2005. How do galaxies get their gas? *MNRAS*, **363**, 2–28.

KHOCHFAR, S. SILK, J.: 2006. A Simple Model for the Size Evolution of Elliptical Galaxies. *ApJ*, **648**, L21–L24.

KOMATSU, E., SMITH, K. M., DUNKLEY, J., BENNETT, C. L., GOLD, B., HINSHAW, G., JAROSIK, N., LARSON, D., NOLTA, M. R., PAGE, L., SPERGEL, D. N., HALPERN, M., HILL, R. S., KOGUT, A., LIMON, M., MEYER, S. S., ODEGARD, N., TUCKER, G. S., WEILAND, J. L., WOLLACK, E., WRIGHT, E. L.: 2011. Seven-year Wilkinson Microwave Anisotropy Probe (WMAP) Observations: Cosmological Interpretation. *ApJS*, **192**, 18.

KOOPMANS, L. V. E., TREU, T., BOLTON, A. S., BURLES, S., MOUSTAKAS, L. A.: 2006. The Sloan Lens ACS Survey. III. The Structure and Formation of Early-Type Galaxies and Their Evolution since $z \sim 1$. *ApJ*, **649**, 599–615.

KORMENDY, J.: 1977. Brightness distributions in compact and normal galaxies. II - Structure parameters of the spheroidal component. *ApJ*, **218**, 333–346.

KORMENDY, J. BENDER, R.: 1996. A Proposed Revision of the Hubble Sequence for Elliptical Galaxies. *ApJ*, **464**, L119.

KORMENDY, J., FISHER, D. B., CORNELL, M. E., BENDER, R.: 2009. Structure and Formation of Elliptical and Spheroidal Galaxies. *ApJS*, **182**, 216–309.

KUIJKEN, K. DUBINSKI, J.: 1994. Lowered Evans Models - Analytic Distribution Functions of Oblate Halo Potentials. *MNRAS*, **269**, 13.

KULL, A., TREUMANN, R. A., BOEHRINGER, H.: 1997. A Note on the Statistical Mechanics of Violent Relaxation of Phase-Space Elements of Dierent Densities. *ApJ*, **484**, 58–+.

LAUER, T. R., FABER, S. M., GEBHARDT, K., RICHSTONE, D., TREMAINE, S., AJHAR, E. A., ALLER, M. C., BENDER, R., DRESSLER, A., FILIPPENKO, A. V., GREEN, R., GRILLMAIR, C. J., HO, L. C., KORMENDY, J., MAGORRIAN, J., PINKNEY, J., SIOPIS, C.: 2005. The Centers of Early-Type Galaxies with Hubble Space Telescope. V. New WFPC2 Photometry. *AJ*, **129**, 2138–2185.

LONGHETTI, M., SARACCO, P., SEVERGNINI, P., DELLA CECA, R., MANNUCCI, F., BENDER, R., DRORY, N., FEULNER, G., HOPP, U.: 2007. The Kormendy relation of massive elliptical galaxies at z $\sim$ 1.5: evidence for size evolution. *MNRAS*, **374**, 614–626.

LOTZ, J. M., JONSSON, P., COX, T. J., CROTON, D., PRIMACK, J. R., SOMERVILLE, R. S., STEWART, K.: 2011. The Major and Minor Galaxy Merger Rates at z < 1.5. *ApJ*, **742**, 103.

LYNDEN-BELL, D.: 1967. Statistical mechanics of violent relaxation in stellar systems. *MNRAS*, **136**, 101–+.

MERRITT, D.: 1996. Optimal Smoothing for N-Body Codes. *AJ*, **111**, 2462–+.

MILLER, R. H. SMITH, B. F.: 1980. Galaxy collisions - A preliminary study. *ApJ*, **235**, 421–436.

MISGELD, I. HILKER, M.: 2011. Families of dynamically hot stellar systems over 10 orders of magnitude in mass. *MNRAS*, **414**, 3699–3710.

MO, H., VAN DEN BOSCH, F. C., WHITE, S.: 2010. *Galaxy Formation and Evolution*.

MONAGHAN, J. J. LATTANZIO, J. C.: 1985. A refined particle method for astrophysical problems. *A&A*, **149**, 135–143.

MOSTER, B. P., MACCIÒ, A. V., SOMERVILLE, R. S., JOHANSSON, P. H., NAAB, T.: 2010. Can gas prevent the destruction of thin stellar discs by minor mergers? *MNRAS*, **403**, 1009–1019.

NAAB, T. BURKERT, A.: 2003. Statistical Properties of Collisionless Equal- and Unequal-Mass Merger Remnants of Disk Galaxies. *ApJ*, **597**, 893–906.

NAAB, T., JOHANSSON, P. H., OSTRIKER, J. P.: 2009. Minor Mergers and the Size Evolution of Elliptical Galaxies. *ApJ*, **699**, L178–L182.

NAAB, T., JOHANSSON, P. H., OSTRIKER, J. P., EFSTATHIOU, G.: 2007. Formation of Early-Type Galaxies from Cosmological Initial Conditions. *ApJ*, **658**, 710–720.

NAAB, T. OSTRIKER, J. P.: 2009. Are Disk Galaxies the Progenitors of Giant Ellipticals? *ApJ*, **690**, 1452–1462.

NAAB, T. TRUJILLO, I.: 2006. Surface density profiles of collisionless disc merger remnants. *MNRAS*, **369**, 625–644.

NAKAMURA, T. K.: 2000. Statistical Mechanics of a Collisionless System Based on the Maximum Entropy Principle. *ApJ*, **531**, 739–743.

NAVARRO, J. F., FRENK, C. S., WHITE, S. D. M.: 1997. A Universal Density Profile from Hierarchical Clustering. *ApJ*, **490**, 493–+.

NEGROPONTE, J. WHITE, S. D. M.: 1983. Simulations of mergers between disc-halo galaxies. *MNRAS*, **205**, 1009–1029.

NELSON, A. F., WETZSTEIN, M., NAAB, T.: 2009. Vine-A Numerical Code for Simulating Astrophysical Systems Using Particles. II. Implementation and Performance Characteristics. *ApJS*, **184**, 326–360.

NIPOTI, C., LONDRILLO, P., CIOTTI, L.: 2003. Galaxy merging, the fundamental plane of elliptical galaxies and the M_{BH}-σ_0 relation. *MNRAS*, **342**, 501–512.

NIPOTI, C., TREU, T., AUGER, M. W., BOLTON, A. S.: 2009a. Can Dry Merging Explain the Size Evolution of Early-Type Galaxies? *ApJ*, **706**, L86–L90.

NIPOTI, C., TREU, T., BOLTON, A. S.: 2009b. Dry Mergers and the Formation of Early-Type Galaxies: Constraints from Lensing and Dynamics. *ApJ*, **703**, 1531–1544.

OSER, L., NAAB, T., OSTRIKER, J. P., JOHANSSON, P. H.: 2011. The cosmological size and velocity dispersion evolution of massive early-type galaxies. *ArXiv e-prints*.

OSER, L., OSTRIKER, J. P., NAAB, T., JOHANSSON, P. H., BURKERT, A.: 2010. The Two Phases of Galaxy Formation. *ApJ*, **725**, 2312–2323.

PAHRE, M. A., DE CARVALHO, R. R., DJORGOVSKI, S. G.: 1998. Near-Infrared Imaging of Early-Type Galaxies. IV. The Physical Origins of the Fundamental Plane Scaling Relations. *AJ*, **116**, 1606–1625.

PRUGNIEL, P. SIMIEN, F.: 1997. The fundamental plane of early-type galaxies: non-homology of the spatial structure. *A&A*, **321**, 111–122.

SALPETER, E. E.: 1955. The Luminosity Function and Stellar Evolution. *ApJ*, **121**, 161.

SARACCO, P., LONGHETTI, M., ANDREON, S.: 2009. The population of early-type galaxies at 1 < z < 2 - new clues on their formation and evolution. *MNRAS*, **392**, 718–732.

SCOTT, N., CAPPELLARI, M., DAVIES, R. L., BACON, R., DE ZEEUW, P. T., EMSELLEM, E., FALCÓN-BARROSO, J., KRAJNOVIĆ, D., KUNTSCHNER, H., MCDERMID, R. M., PELETIER, R. F., PIPINO, A., SARZI, M., VAN DEN BOSCH, R. C. E., VAN DE VEN, G., VAN SCHERPENZEEL, E.: 2009. The SAURON Project - XIV. No escape from V_{esc}: a global and local parameter in early-type galaxy evolution. *MNRAS*, **398**, 1835–1857.

SERSIC, J. L.: 1968. *Atlas de galaxias australes*.

SHEN, S., MO, H. J., WHITE, S. D. M., BLANTON, M. R., KAUFFMANN, G., VOGES, W., BRINKMANN, J., CSABAI, I.: 2003. The size distribution of galaxies in the Sloan Digital Sky Survey. *MNRAS*, **343**, 978–994.

SHU, F. H.: 1978. On the statistical mechanics of violent relaxation. *ApJ*, **225**, 83–94.

SPERGEL, D. N., BEAN, R., DORÉ, O., NOLTA, M. R., BENNETT, C. L., DUNKLEY, J., HINSHAW, G., JAROSIK, N., KOMATSU, E., PAGE, L., PEIRIS, H. V., VERDE, L., HALPERN, M., HILL, R. S., KOGUT, A., LIMON, M., MEYER, S. S., ODEGARD, N., TUCKER, G. S., WEILAND, J. L., WOLLACK, E., WRIGHT, E. L.: 2007. Three-Year Wilkinson Microwave Anisotropy Probe (WMAP) Observations: Implications for Cosmology. *ApJS*, **170**, 377–408.

SPERGEL, D. N. HERNQUIST, L.: 1992. Statistical mechanics of violent relaxation. *ApJ*, **397**, L75–L78.

SPRINGEL, V.: 2005. The cosmological simulation code GADGET-2. *MNRAS*, **364**, 1105–1134.

SPRINGEL, V., WHITE, S. D. M., JENKINS, A., FRENK, C. S., YOSHIDA, N., GAO, L., NAVARRO, J., THACKER, R., CROTON, D., HELLY, J., PEACOCK, J. A., COLE, S., THOMAS, P., COUCHMAN, H., EVRARD, A., COLBERG, J., PEARCE, F.: 2005. Simulations of the formation, evolution and clustering of galaxies and quasars. *Nature*, **435**, 629–636.

STIAVELLI, M. BERTIN, G.: 1987. Statistical mechanics and equilibrium sequences of ellipticals. *MNRAS*, **229**, 61–71.

SZOMORU, D., FRANX, M., VAN DOKKUM, P. G.: 2011. Sizes and surface brightness profiles of quiescent galaxies at z ~ 2. *ArXiv e-prints*.

TAYLOR, E. N., FRANX, M., GLAZEBROOK, K., BRINCHMANN, J., VAN DER WEL, A., VAN DOKKUM, P. G.: 2010. On the Dearth of Compact, Massive, Red Sequence Galaxies in the Local Universe. *ApJ*, **720**, 723–741.

TOFT, S., VAN DOKKUM, P., FRANX, M., LABBE, I., FÖRSTER SCHREIBER, N. M., WUYTS, S., WEBB, T., RUDNICK, G., ZIRM, A., KRIEK, M., VAN DER WERF, P., BLAKESLEE, J. P., ILLINGWORTH, G., RIX, H.-W., PAPOVICH, C., MOORWOOD, A.: 2007. Hubble Space Telescope and Spitzer Imaging of Red and Blue Galaxies at z ~ 2.5: A Correlation between Size and Star Formation Activity from Compact Quiescent Galaxies to Extended Star-forming Galaxies. *ApJ*, **671**, 285–302.

TOOMRE, A.: 1977. Mergers and Some Consequences. In B. M. Tinsley & R. B. G. Larson D. Campbell, editor, *Evolution of Galaxies and Stellar Populations*, page 401.

TOOMRE, A. TOOMRE, J.: 1972. Galactic Bridges and Tails. *ApJ*, **178**, 623–666.

TREMAINE, S., RICHSTONE, D. O., BYUN, Y., DRESSLER, A., FABER, S. M., GRILLMAIR, C., KORMENDY, J., LAUER, T. R.: 1994. A family of models for spherical stellar systems. *AJ*, **107**, 634–644.

TRUJILLO, I., ASENSIO RAMOS, A., RUBI ~NO-MARTÍN, J. A., GRAHAM, A. W., AGUERRI, J. A. L., CEPA, J., GUTIÉRREZ, C. M.: 2002. Triaxial stellar systems following the $r^{1/n}$ luminosity law: an analytical mass-density expression, gravitational torques and the bulge/disc interplay. *MNRAS*, **333**, 510–516.

TRUJILLO, I., CENARRO, A. J., DE LORENZO-CÁCERES, A., VAZDEKIS, A., DE LA ROSA, I. G., CAVA, A.: 2009. Superdense Massive Galaxies in the Nearby Universe. *ApJ*, **692**, L118–L122.

TRUJILLO, I., CONSELICE, C. J., BUNDY, K., COOPER, M. C., EISENHARDT, P., ELLIS, R. S.: 2007. Strong size evolution of the most massive galaxies since z ~ 2. *MNRAS*, **382**, 109–120.

TRUJILLO, I., ERWIN, P., ASENSIO RAMOS, A., GRAHAM, A. W.: 2004. Evidence for a New Elliptical-Galaxy Paradigm: Sérsic and Core Galaxies. *AJ*, **127**, 1917–1942.

TRUJILLO, I., FERRERAS, I., DE LA ROSA, I. G.: 2011. Dissecting the size evolution of elliptical galaxies since z ~ 1: pu ng-up versus minor-merging scenarios. *MNRAS*, **415**, 3903–3913.

TRUJILLO, I., FÖRSTER SCHREIBER, N. M., RUDNICK, G., BARDEN, M., FRANX, M., RIX, H.-W., CALDWELL, J. A. R., MCINTOSH, D. H., TOFT, S., HÄUSSLER, B., ZIRM, A., VAN DOKKUM, P. G., LABBÉ, I.: 2006. The Size Evolution of Galaxies since z ~3: Combining SDSS, GEMS, and FIRES. *ApJ*, **650**, 18–41.

TRUJILLO, I., GRAHAM, A. W., CAON, N.: 2001. On the estimation of galaxy structural parameters: the Sérsic model. *MNRAS*, **326**, 869–876.

VAN ALBADA, T. S. VAN GORKOM, J. H.: 1977. Experimental Stellar Dynamics for Systems with Axial Symmetry. *A&A*, **54**, 121.

VAN DE SANDE, J., KRIEK, M., FRANX, M., VAN DOKKUM, P. G., BEZANSON, R., WHITAKER, K. E., BRAMMER, G., LABBÉ, I., GROOT, P. J., KAPER, L.: 2011. The Stellar Velocity Dispersion of a Compact Massive Galaxy at z = 1.80 Using X-Shooter: Confirmation of the Evolution in the Mass-Size and Mass-Dispersion Relations. *ApJ*, **736**, L9.

VAN DER WEL, A., FRANX, M., VAN DOKKUM, P. G., RIX, H.-W., ILLINGWORTH, G. D., ROSATI, P.: 2005. Mass-to-Light Ratios of Field Early-Type Galaxies at z ~ 1 from Ultradeep Spectroscopy: Evidence for Mass-dependent Evolution. *ApJ*, **631**, 145–162.

VAN DER WEL, A., HOLDEN, B. P., ZIRM, A. W., FRANX, M., RETTURA, A., ILLINGWORTH, G. D., FORD, H. C.: 2008. Recent Structural Evolution of Early-Type Galaxies: Size Growth from z = 1 to z = 0. *ApJ*, **688**, 48–58.

VAN DOKKUM, P. G., FRANX, M., KRIEK, M., HOLDEN, B., ILLINGWORTH, G. D., MAGEE, D., BOUWENS, R., MARCHESINI, D., QUADRI, R., RUDNICK, G., TAYLOR, E. N., TOFT, S.: 2008. Confirmation of the Remarkable Compactness of Massive Quiescent Galaxies at z ~ 2.3: Early-Type Galaxies Did not Form in a Simple Monolithic Collapse. *ApJ*, **677**, L5–L8.

VAN DOKKUM, P. G., KRIEK, M., FRANX, M.: 2009. A high stellar velocity dispersion for a compact massive galaxy at redshift z = 2.186. *Nature*, **460**, 717–719.

VAN DOKKUM, P. G., WHITAKER, K. E., BRAMMER, G., FRANX, M., KRIEK, M., LABBÉ, I., MARCHESINI, D., QUADRI, R., BEZANSON, R., ILLINGWORTH, G. D., MUZZIN, A., RUDNICK, G., TAL, T., WAKE, D.: 2010. The Growth of Massive Galaxies Since z = 2. *ApJ*, **709**, 1018–1041.

VILLUMSEN, J. V.: 1982. Simulations of galaxy mergers. *MNRAS*, **199**, 493–516.

VILLUMSEN, J. V.: 1983. Simulations of galaxy mergers. II. *MNRAS*, **204**, 219–236.

WAKE, D. A., WHITAKER, K. E., LABBÉ, I., VAN DOKKUM, P. G., FRANX, M., QUADRI, R., BRAMMER, G., KRIEK, M., LUNDGREN, B. F., MARCHESINI, D., MUZZIN, A.: 2011. Galaxy Clustering in the NEWFIRM Medium Band Survey:

The Relationship Between Stellar Mass and Dark Matter Halo Mass at $1 < z < 2$. *ApJ*, **728**, 46.

WETZSTEIN, M., NELSON, A. F., NAAB, T., BURKERT, A.: 2009. Vine-A Numerical Code for Simulating Astrophysical Systems Using Particles. I. Description of the Physics and the Numerical Methods. *ApJS*, **184**, 298–325.

WHITE, S. D. M.: 1978. Simulations of merging galaxies. *MNRAS*, **184**, 185–203.

WHITE, S. D. M.: 1979. Further simulations of merging galaxies. *MNRAS*, **189**, 831–852.

WHITE, S. D. M.: 1980. Mixing processes in galaxy mergers. *MNRAS*, **191**, 1P–4P.

WHITE, S. D. M. REES, M. J.: 1978. Core condensation in heavy halos - A two-stage theory for galaxy formation and clustering. *MNRAS*, **183**, 341–358.

WILLIAMS, R. J., QUADRI, R. F., FRANX, M.: 2011. The Diminishing Importance of Major Galaxy Mergers at Higher Redshifts. *ApJ*, **738**, L25.

WILLIAMS, R. J., QUADRI, R. F., FRANX, M., VAN DOKKUM, P., TOFT, S., KRIEK, M., LABBÉ, I.: 2010. The Evolving Relations Between Size, Mass, Surface Density, and Star Formation in 3×10^4 Galaxies Since $z = 2$. *ApJ*, **713**, 738–750.

WUYTS, S., COX, T. J., HAYWARD, C. C., FRANX, M., HERNQUIST, L., HOPKINS, P. F., JONSSON, P., VAN DOKKUM, P. G.: 2010. On Sizes, Kinematics, M/L Gradients, and Light Profiles of Massive Compact Galaxies at $z \sim 2$. *ApJ*, **722**, 1666–1684.

ZIRM, A. W., VAN DER WEL, A., FRANX, M., LABBÉ, I., TRUJILLO, I., VAN DOKKUM, P., TOFT, S., DADDI, E., RUDNICK, G., RIX, H., RÖTTGERING, H. J. A., VAN DER WERF, P.: 2007. NICMOS Imaging of DRGs in the HDF-S: A Relation between Star Formation and Size at $z \sim 2.5$. *ApJ*, **656**, 66–72.

LIST OF PUBLICATIONS

- **Hilz, M.**, Naab T. & Ostriker J.P. *"How do minor mergers promote inside-out growth of ellipticals, transforming the size, density profile and dark matter fraction?"*, accepted by MNRAS

- **Hilz, M.**, Naab, T., Ostriker J.P. et al. *"Relaxation and Stripping: The Evolution of Sizes and Dark Matter Fractions in Major and Minor Mergers of Elliptical Galaxies"*, published in MNRAS

i want morebooks!

Buy your books fast and straightforward online - at one of world's fastest growing online book stores! Environmentally sound due to Print-on-Demand technologies.

Buy your books online at
www.get-morebooks.com

Kaufen Sie Ihre Bücher schnell und unkompliziert online – auf einer der am schnellsten wachsenden Buchhandelsplattformen weltweit! Dank Print-On-Demand umwelt- und ressourcenschonend produziert.

Bücher schneller online kaufen
www.morebooks.de

 VDM Verlagsservicegesellschaft mbH
Heinrich-Böcking-Str. 6-8   Telefon: +49 681 3720 174   info@vdm-vsg.de
D - 66121 Saarbrücken   Telefax: +49 681 3720 1749   www.vdm-vsg.de

Printed by Books on Demand GmbH, Norderstedt / Germany